BIRD-WITCHED

Bird World symbol, designed by Marjorie Valentine Adams

BIRD-WITCHED

How Birds Can Change a Life

MARJORIE VALENTINE ADAMS
FOREWORD BY GREG LASLEY AND CHUCK SEXTON

UNIVERSITY OF TEXAS PRESS
Austin

Mildred Wyatt-Wold Series in Ornithology

Requests for permission to reproduce material from this work
should be sent to:
Permissions
University of Texas Press
P.O. Box 7819
Austin, TX 78713-7819
www.utexas.edu/utpress/about/bpermission.html

∞ The paper used in this book meets the minimum requirements of
ANSI/NISO Z39.48-1992 (R1997) (Permanence of Paper).
Library of Congress Cataloging-in-Publication Data
Adams, Marjorie Valentine, 1913–
Bird-witched : how birds can change a life / Marjorie Valentine Adams ;
foreword by Greg Lasley and Chuck Sexton. — 1st ed.
p.   cm. — (Mildred Wyatt-Wold series in ornithology)
ISBN 978-0-292-71925-5
1. Bird watching — Texas — Anecdotes.   2. Adams, Marjorie
Valentine, 1913–   I. Title.   II. Series.
QL684.T4A3199 2005
598′.072′34 — dc22
2005007854

Dedicated to and in Loving Memory of
Louis Taft "Red" Adams
1911–2005
I couldn't have done it without you.
(Portrait by Miles Mathis)

And to Lew and Louise Adams,
pillars of support

*Marjorie Adams—Award-winning writer, filmmaker,
syndicated columnist, and conservationist.
(Publicity photo, furnished to editors circa 1960s)*

CONTENTS

FOREWORD

Greg Lasley and Chuck Sexton

The two of us came to the Texas birding scene some thirty years ago. Birders and nature filmmakers Marjorie and Red Adams were already among the elite of Texas naturalists of the day, and anyone learning about birds in Texas was sure to make their acquaintance sooner or later. However, our introductions came from divergent pathways. Greg Lasley and his wife met Marjorie and Red at an Audubon gathering. Greg was interested in Red's photography, and soon Marjorie and Red took the couple under their wings to guide them on some of their early birding explorations of Texas bird life. The chase was on.

At the same time, Chuck Sexton was a graduate student at the University of Texas in Austin. He was a birder, to be sure, but nonetheless focused on academic pursuits with a keen interest in the impacts of humankind on bird populations. On a serendipitous trip to nearby McKinney Falls State Park, Chuck stopped in the visitors center and viewed a short film produced by the Adams duo and titled *What Good Is a Warbler?* Using the life of a single endangered bird species, the Golden-cheeked Warbler, the award-winning film was becoming an enlightening lesson for the public and for schoolchildren far beyond Texas borders, even as far away as Australia. The film immediately helped solidify Chuck's concept of his graduate research studies and his professional career goals—the conservation of rare species in rapidly urbanizing landscapes.

From this combination of sport and academia, we both came under the spell of this charming duo. Now, we find both the thrills of sport birding and the struggles of conservation efforts richly portrayed in Marjorie's book, *Bird-Witched: How Birds Can Change a Life.*

This is no simple linear chronology of birding events or of life in general, starting in the rugged Texas Hill Country, where there's not much linearity in the land nor in its pathway to the present. We share Marjorie and Red's familiarity with these haunts, and, for us, some of her stories of comfortingly familiar scenes and creatures flow out like one of the thousands of sparkling springs in this distinctive and multifaceted landscape. To the uninitiated, the stories offer a window to the area's amazing biological diversity.

Yet the Texas Hill Country forms only the home base for the array of other adventures in this book. From the day more than forty years ago when a Red-bellied Woodpecker grabbed Marjorie's attention, the Adams pair traveled from Canada to Mexico and Belize and along the way searched out not only birds but the bird people, both noted professionals as well as amateurs, who know birds and the land on which the birds' lives depend. These people included Roger Tory Peterson, Roy Bedichek, Edgar Kincaid Jr., Victor Emanuel, Dr. Alvarez del Toro, Dr. Charles Hartshorne, Connie Hagar, Irby Davis, Jim Tucker, and a host of others.

The myriad anecdotes that flow from Marjorie's computer give the sense of the passage of history: Roger Peterson and his sight of the Ivory-billed Woodpecker. Victor Emanuel and the Eskimo Curlew. The first record of the Green Violet-ear in the United States. The last California Condor in the wild. The battle for the Golden-cheeked Warbler. The birding world as it unfolded and evolved into the passion it has become today.

For us, *Bird-Witched* offers an ecology lesson in nearly every adventure. Whether the lessons are whimsical nature observations from the Throne or gleaned from the aforementioned luminaries, they flow forth as an earnest search for meaning and results from the pursuit and study of birds and the lands on which they live.

Marjorie often refers to herself as "Old Lady of Texas," and indeed the anecdotes that flow from her computer provide a rich sense of history: the remembrances of someone who was a part of the birding world as it evolved to today's popular pursuit. She not only commented on the conservation issues of her time but also taught and inspired a generation of others to move those conservation goals forward into the future. With *Bird-Witched* we have a front-row seat to birds and birding for the past forty years, as seen through the eyes of a caring and thinking guardian of Nature.

We are honored that Marjorie calls us her friends. Old Lady of Texas? Hardly. To us Marjorie is a Grand Lady of Birds—and always will be.

ACKNOWLEDGMENTS

Fate has arranged for me to outlive some of those involved in the adventures and information in this lifetime book. I pause here to honor their memories.

I am also grateful to Rob Fergus, founding director of the Hornsby Bend Bird Observatory, who was official First Reader and spent hours analyzing my manuscript in careful detail. Thank you, Rob, and I tried to follow every suggestion you made.

Father Tom Pincelli, as Second Reader, evaluated the manuscript not only from a birder's viewpoint but also as a tool for teaching. I am grateful for your praise of me as a senior still working to make the world a little better.

Thank you, Rosemary Wetherold, for your gentle, knowledgeable editing. It's a pity I didn't have you years ago.

I especially want to remember here Shannon Davies, who not only took me birding and around and about many times but also helped me realize I might have a book inside me.

John Kelly, who helped me as if I was his own grandmother, was especially helpful with the complicated history of Hornsby Bend, and he even delivered the manuscript to UT Press.

Let's not forget the American Birding Association and Nanci Hawley, ABA meetings manager, who furnished the only known pictures of ABA's first convention as well as much other information.

And here again, thanks for the constant help of Lew and Louise Adams every step of the way.

Good birding to all,
Old Lady of Texas

BIRD-WITCHED

PROLOGUE

How It Began

.

Never underestimate the power of birds.

Dig as deep as I can into the past, I can't recall when I formally began to play the game and sport now called birding, but I do have vivid memories of how my husband, Red, and I upset our lives more than forty years ago. I gave him binoculars and he gave me Roger Tory Peterson's new book, *A Field Guide to the Birds of Texas.*

Out in our yard a bird was pecking a hole in a tree. When I focused the binoculars on it, something happened without warning. I stared at this beautifully designed work of art, with my brain cells working furiously, and the result was nothing less than complete awe. This extraordinary creature resided just outside my door, and I had never given it a glance, much less given thought to woodpecker real estate rights. Our new guide identified it as a Red-bellied Woodpecker.

When Red found some gawky neighbors in the park below our house, we matched them to the picture of the little Green Heron. Only thirteen city blocks from the Texas state capitol we watched incredulously as a nesting Greater Roadrunner stuffed an outsized lizard down its baby's throat. With the lizard's back legs and tail wiggling fiercely, baby was sure to choke, but baby sat steadfastly at attention and the lizard slid down slowly as it was digested.

As I worked at my desk, a pair of Carolina Chickadees just outside my window showed me in the most loving way that birds can be tender.

A late spring norther blustered south, and in the morning bright flashes of color moved rapidly through the trees to refuel on insects. Spotted forms rested unobtrusively under hedges, then raced across the lawn with outspread wings to pounce on small prey. From a thicket a heavenly melody ascended. Could such a harmony come from a "lowly" creature?

A strange mania grabbed me. All I wanted to do was see the creatures that were doing all these things. All I wanted to do was chase birds. Housework, children, and chores became obstacles to what was fast becoming the most compelling thing in my life. *It was awful!*

First and foremost it was the myriad complexes of avian beauty, design, and voice; then it was the mystery of the birds' lives in a world I had never experienced. Finally, the dread primeval hunting instinct took possession. As I unknowingly followed in the footsteps of the great John James Audubon, it became the utmost necessity that I find and see every kind of bird as expeditiously as possible to make it mine, mine, mine. I had "caught" it. I was bird-witched!

I am not the first person to thus lose sanity, and I have witnessed a series of these other fortunate souls through the years. The first was a young student, desperately needing someone, anyone, to talk to. During the peak of bird migration he had gone "crazy-wild" over birds, suffered severe indigestion due to excitement, dreamed about birds every night, and scarcely could wait for dawn so that he could spend every free moment outdoors. All else in his life had suddenly become secondary. He was having a difficult time with friends and relatives who openly let him know their judicious opinion that birdwatching was a pastime and unconscionable as a career pursuit.

"But I can't help it," he confessed. Then he almost moaned, "I don't care if I starve and I don't have a single friend or any family left. I'm going to study ornithology."

I didn't tell him there was not a single university or college that offered a major solely in that discipline. The sad fact is that this remains true today.

Like this fortunate young man, I too went crazy-wild and decided that birds should be my life work, but it was not always thus. In fact, in my young days birds and I got off to an ornery start.

The problem began on my great-uncle Walter's ranch, about fifty miles from San Antonio. My duties were setting the table, drying dishes, churning butter, and sweeping the wide veranda and the steps, each of which was a single five-foot-long block of limestone.

After that, what else could I do for entertainment?

Both aunt and uncle were widowed and without children, and it probably never occurred to them that it could be dangerous for an eleven-year-old city girl to wander alone anywhere and everywhere in the rugged countryside that my skinny legs could carry me. Of course, I had the three dogs with me and I was armed with the stout stalk of a twisted-leaf yucca. I can still remember the thrill of raising this humble scepter on high as I stood triumphantly on top of the tallest hill and declared the entire spread to be my kingdom.

The racing stallion, Texas Ranger, with Jim Valentine, Marjorie's brother, astride and great-uncle Walter and Marjorie standing by at the barn, Bonita Ranch, Texas. (Photo by Wilma Davis Valentine)

There is a limit, however, to exploration, to tree climbing in the giant live oaks in the yard, and to cleaning ticks and fleas off the dogs. In that ancient time there was no radio and no TV, and the bookcase in the hall offered to my taste only Zane Grey's famous western *Riders of the Purple Sage*.

I resorted to our only tie to the outside world: the telephone that hung on the wall in an oaken case in the dining room. It seldom gave the number of rings for the Askey household, but never mind. I began listening in on the party line. My punishment for invading privacy? All the conversations were in German.

Thus it fell to the chickens to keep me occupied. I reveled in gathering the eggs, a glorious contest between me and each hen. The barn, with its various stables, storerooms, and huge hayloft, seemed a great castle, and like all great castles, it had its forbidden danger. I was never under any circumstance to go near Texas Ranger, Uncle Walter's greatest pride—a powerful roan stallion, a prince of horses, and a winner on the racetrack. A lowly goat shared his stall to keep him calm.

The hens made use of the whole barn, and it was a daily challenge to prove I was smarter than any wily hen that compacted her feathers in a disappearing act and at any moment could catapult into the air, powered by a din of furious cackles and chicken insults.

The chickens were not without revenge. When I brought a baby chicken home with me and cuddled the fluffy darling close to love it, it returned the favor with an excruciating peck in my eye.

The next unfortunate incident with a bird was a sudden, furious bomb attack that left my forehead bleeding. How could I know I had gotten too close to important things in a mockingbird's territory?

Finally, it was a bird that reduced me to the level of a prokaryotic slime and, even worse, a drunk one. Some stories are so odd they are bound to be lies, but it is absolutely true that birds can range over large territories. It is also true that their various activities can take place in random spots. It was a billion-to-one quirk that out in the schoolyard a bird flew over, aimed, and plopped a large nitrogen-rich gift on my head. That was the day I learned I didn't have a friend in the entire school.

But now, years later, the power of birds easily erased this inglorious ornithological past. Armed with their magic, I winged my way to ecstasy, and it seemed natural for enchantment to progress to inspiration. I had been writing since age twelve, when my first story was published in the school newsletter, and I had been publishing since 1935. So, of course, I now would write about birds. As the months passed, the decision came: I would create a feature about birds for all the newspapers in the country. With saintly generosity I would share the delight of birds with the waiting public, and Bird World would be an instant hit. How could it be otherwise? For no living being anywhere on the planet could resist the power of birds.

Incidentally, it was possible I might also make a fortune. It didn't occur to me that the formula "birds = $$" was decades into the future, and the pittance that newspapers paid was well known.

In fact, there were several minor conditions that never weighed on my brain. Foremost, we were in the predawn of the age of environmental awareness. The term "environmentalist" or "conservationist" or even "natural world" was not included in my dictionary or thesaurus. DDT and other persistent chemical compounds were being used as miracles of science. The required environmental impact statement was not even an embryo, and the job description "nongame biologist" was as rare as the proverbial hen's teeth.

There was no Greenpeace USA, United Nations Environment Program, HawkWatch International, Waterfowl USA, or Friends of the Earth. No Earth First!, Environmental Defense Fund, Rachel Carson Memorial Fund, National Park Service, or U.S. Fish and Wildlife Service. No America the Beautiful Fund, Earth Day, or Bureau of Land Management. No Environmental Protection Agency, Union of Concerned Scientists, or Natural Resources

Defense Council. No Clean Air Act, Clean Water Act, Endangered Species Act, or CITES (Convention on International Trade in Endangered Species of Wild Fauna and Flora). There was not even a Ranger Rick Nature Club. Even worse, there was no American Birding Association, and birding, recognized nationally and internationally today as a game and sport with standard rules and ethics, was in the delivery room but still awaiting birth.

Another minor problem was that I didn't know much about birds.

My first bird experience had happened in elementary school. In the 1920s the city of Austin, Texas, boasted towering streetlights 165 feet tall. There had been a severe storm in the night, and the next morning below one of these lights we schoolchildren discovered dozens of dead birds. Our teacher gathered the little bodies in several boxes and told us that in the storm the birds had been attracted to the light and flew into it by mistake. I don't know if she knew anything about birds, but she appreciated them enough to mourn they had died, and she let us hold them in our hands so we could see the feather patterns and the many colors, which varied from bird to bird. It was a close encounter with feathers, the clothes birds wear . . . a close encounter of the bird kind.

In San Antonio my formal scientific education consisted of one class in junior high school in which our well-meaning teacher not only knew nothing about birds but also was probably teaching because the originally contracted teacher had managed to get married. All I can remember about science from this class involved the wildflowers I picked on the way to school, which were then duly identified with the current local name. The situation was somewhat better in math, English, and especially Texas history classes, which occurred in multiples. It's a sad story, but that really was school in my neighborhood in San Antonio, Texas, almost eighty years ago.

In college my interests were art and English, and I never stepped foot in the science building. And husband Red's schooling in East Texas was even worse. However, I recently came across a small brown envelope, postage one cent, addressed to me where I lived at about age twelve. Inside were some Arm and Hammer bird cards, so I must have had some interest in avian creatures even then.

As an adult, I had been studying birds sporadically for about ten years, using a book with pen-and-ink drawings of birds, some posed on their backs with feet in air, and all accompanied by wearisome scientific descriptions. I had also published an article in 1956 about the beloved Texas naturalist Roy Bedichek, titled "Bedi the Bird Man," and I was grateful for occasional advice from him.

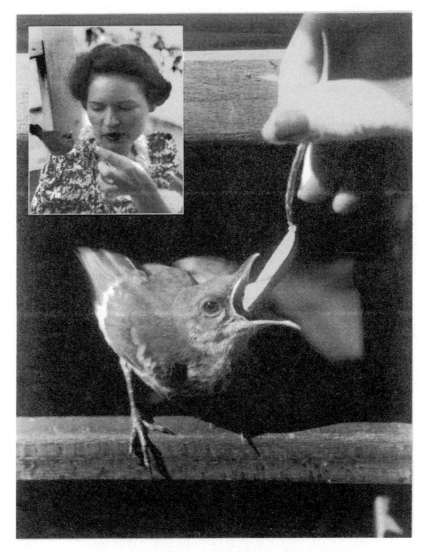

Marjorie rescues and becomes a mother to a baby mockingbird.
(Photo by Marjorie Adams, published in Look *magazine. Inset photo by Red Adams)*

But never mind my impoverished ornithological education. Roger Tory Peterson had opened the jailhouse door by publishing his field guides to birds. With exhilarating freedom I could find and know birds anywhere on my own.

Was it good fortune that I didn't realize how ignorant I was as I blithely proceeded to chase the living bird and do research anywhere and everywhere? Without doubt, it was good fortune for me that Red was an excellent field man who not only was willing to tolerate all this investigation but also was in his element as we shared my outdoor bird search. He had been born in the fabled Big Thicket of East Texas, had begun hunting as soon as he could drag a gun behind him, had cleverly avoided snakebite as he fished all the creeks, and had always been able to find his way home in that wilderness. Best of all, he had a keen ear and eye for birds.

In fact, Red is something of an original. Not long ago my sister-in-law, Mary Alice Valentine, said that when she came to Texas from New York in 1943, she was fascinated with Red's deep–East Texas lingo. "He never uses a cliché," she noted. "For instance, when I asked him what near beer tasted like, his answer was 'Like stump water with a bar of soap in it.'"

It was natural for Red to praise someone by saying, "You make me as proud as a stud cricket," or to note that an event happened "quicker than rain" or "wouldn't that beat a goose a-gobbling?" When we got tired, we were "running out of Cosmoline," a jelly used to clean guns. Red would commiserate by saying, "That's worse than Ned in the first reader," or he might remark on a delicious meal by saying that it "sure was a lot better than cold coon and collards." He might note that he could do a better job than the one just performed even if he had to do it "on a pair of Tom-walkers" (stilts). Soda pop was "belly wash." Red could recite long chants that began with "Born in the backwoods, suckled by a bear" or with "This is the tale of Big-bellied Ben; he could eat more meat than forty men." Red is fun.

He had his own names for birds. For instance, a Pileated Woodpecker was a "Good God!" because the bird's call was so loud that anyone who heard it would exclaim, "Good God! What was that?" A Summer Tanager was a "bee-eater." Also young Red was such a good shot that he had killed objects as small as flying wasps with a forked slingshot, a weapon called at that time by a name that is now extirpated. Red was born to be a woodsman. How lucky I have been to have him as a partner!

It was Red who won a trip to Florida, and while we were in Tavernier on the Florida Keys, we looked up Sandy Sprunt, research director of the National Audubon Society. Not only did that learned birdman tell us where we could find a Swallow-tailed Kite nest, but he also agreed to give Bird World the once-over. I had put together a sample thirteen-week supply of columns, and when he returned them, he wrote that they were accurate and should be well accepted by the reading public. His only suggestion: it possibly would be better to state that there are more Wild Turkeys today "than before the white man came."

Back home, I handed the same packet to Charles Green, editor of the *Austin American-Statesman,* our local newspaper, humbly asking him, "Could you please do me the favor of looking these over and giving me some advice on what to do with them?"

The good editor pointed me to a chair, then disappeared into his office. When he came out, he exclaimed, "Marjorie, I'll buy these!"

I stood there, trying to figure out what he had just said. Was it "I'll buy these"? I made every effort to look nonchalant. By setting my jaw and knuckling my fists, I managed not to hug this man I barely knew who so suddenly was changing my life more than either of us could imagine.

Soaring with encouragement, I worked with an artist to design my column logo, I had Bird World columns mimeographed into many copies, collated them by hand into piles of packets with letter, résumé, and other goodies, and mailed them to the editors of most of the better newspapers. It was solid work, considerable expense, and a total failure.

It wasn't until I began calling on editors in person that I began to see the real world. My father for a good part of his life had been a traveling salesman, so I grew up knowing that selling is a challenging job. I had done some selling myself—things like paintings or mail-order dresses during World War II, and then John Connally and Jake Pickle and Ed Syers had taught me additional lessons as we all sold advertising at radio station KVET in Austin. I had even rescued a statewide trade magazine for Texas homebuilders by not only editing it but also selling advertising for it.

The trouble was that birdwatching had yet to come out of the closet where the general public traditionally ensconced it, along with little old ladies in tennis shoes, tweedy matrons perched in trees who screeched and squabbled and were accompanied only by men of little muscle. The proposition that it could be a game, much less a sport, was vastly far-fetched, especially for a male (and all the editors were males) and especially in Texas, where, as the trite old saying goes, men are men and women are glad of it.

But on the strength of Charlie Green's purchase, Harry Provence, editor of the *Waco Tribune-Herald,* gave Bird World a fling. Then when I walked into Rhea Howard's office at the *Wichita Falls Times,* I was startled to find a man who loved birds. Not only was this miraculous and my good fortune, but Howard had also just gotten a letter from President Lyndon Johnson, thanking him for the bird feeder Howard had given Lady Bird on his last visit to the ranch. It seemed the letter also included a growl: "I'm not going to feed all those damned sparrows!" However, it was obvious that Lady Bird was keeping the feeder going full swing.

I was in business. I had three newspapers, and I was going to get a total of approximately $105 a month for it. I also had gotten myself in an excellent position to be named "Birdbrained of the Year." I never recovered, never got

Marjorie and Red, screening one of their films at the Times *building in Wichita Falls
to encourage the formation of the North Texas Bird and Nature Club.*
(Courtesy of Wichita Falls Times*)*

smart—just kept going on, maybe getting a $2 raise, maybe getting another
paper, then losing a paper, then getting another. Worst of all, I was faced with
losing my own hometown paper and would have if I hadn't received thirty-
eight letters that week (one of them from Canada), which proved to the new
features editor that I actually had readers.

Red and I had changed our lifestyle from a four-bedroom home to a
twenty-two-foot travel trailer, and for business and for family reasons we traveled from Florida to California. Everywhere we went, we tried to arrange our
stops at refuges, parks, or some nature-oriented spot. Everywhere we went,
I tried to find "bird people" by inquiring at public libraries, zoos, refuges,
museums, and the like. We were amazed and delighted by, and beholden to,
people we had just met for whom we could feel an instant empathy. Sometimes a person was so starved for "bird" fellowship that he or she pleaded with
us to stay just a little longer. Many people made special efforts to show us their
local bird specialties, and if you are one of those and reading this page, I thank
you from my heart again and again. You birders truly are exceptional folks.

It seemed fate that about this time we met James A. Tucker, the father-to-be of the American Birding Association. Red had begun chasing birds with a 16 mm camera and now had enough movie footage for us to attempt a basic film about birding as a game and a sport. We were making a presentation on this theme for the Travis Audubon Society (in the Austin area), and Jim was there, having recently moved from Florida to Texas to work on his PhD at the University of Texas. Jim had invented a book called a *Combination List and Checklist for Birds of North America,* which, by combining alternate full and half-pages, allowed a bird to be listed across the pages not only as a "Lifer" (a bird seen and identified for the first time in the birder's life) but also in the various states. We showed this ingenious book in our film as we checked off a bird in it.

After the program, Jim introduced himself as the book's inventor. As our friendship progressed, we learned that he had begun bird walks in his mother's womb, that birds continued to be a major part of his life, and that, like us, he wanted to make them more so.

So one day we received a journal called *Bird Watcher's Digest,* volume 0, number 0, proclaiming itself to be "a journal devoted to the hobby of bird-watching." It was four pages long, beautifully duplicated in purple ink by hectograph. It included an editorial titled "What Is a Birdwatcher?" and a page of "Games Birdwatchers Play," which described three general types of games that involved making different lists of the birds that the birdwatcher had seen: the Life List (a list of all the birds seen in a person's lifetime), the American Ornithologists' Union (AOU) Territory List, and the World List. There was also the Annual List (all birds seen during a calendar year within the AOU area) and State Lists (all species seen within the confines of a state, excluding Hawaii).

Only nine birders had submitted AOU lists, and the same birders were also the ones who had submitted most of the State Lists. No Life List totals were published in this first issue of the journal. There was also a half-page entitled "Bird Finding News," about the Mexican Crow's invasion of Texas, the second nesting of the Masked Duck in Texas, and the discovery of a nesting Jacana near Kingsville.

The journal promised that rules for each listing game would be forthcoming, but for the present "we would be interested in hearing what each of you feels the rules should be for each game." The subscription rate for the journal was $3 per year and would include a year's membership in the American Birdwatchers Association. A life membership was $50. This section ended with an invitation to the reader to become a member of the association "at the suggestion of myself." It was signed by Jim Tucker.

Looking back, I think that if I had any part at all in forming what is now the official birding game, it was my steadfast use of the terms "birding" and "birder" in my column Bird World, in all of our programs, in all of my birding classes, and in our general conversation. To overcome the stigma attached to the term "birdwatcher," others such as world birder Stuart Keith also decided these terms were more appropriate, and Jim Tucker changed the organization's name to American Birding Association.

However, it took a while. I found in my files a letter I wrote in 1968 to my editor at the *Odessa American* protesting the use of the term "birdies" by a UPI writer. Indeed, as late as 1975, I was debating the term "bird listing" as the name of the game with one of my editors at *Reader's Digest*.

I recall most strongly an incident bordering on the dangerous that involved the term "birder." I was one of the first women to be accepted into the Texas Outdoor Writers Association (TOWA). At my first attendance at a convention, a member was addressing the group and used the word "birdwatcher" in a derogatory tone to denote a group supposedly opposed to hunting and other truly manly sports. Without thinking, I interrupted from the back of the room: "Don't call us birdwatchers. We're *birders*."

Every head turned, wearing the expression, "How the [unprintable] did this [unprintable] female get in here, and what the [unprintable] is she blabbering about?"

It took a while, but the happy ending was that as time passed, I was the recipient of wonderful tales from these same sportsmen regarding their own experiences with birds in their varied outdoor lives. In fact, I have lived long enough to see a female president of TOWA.

For Jim Tucker I was able to publicize in Bird World the brand-new American Birding Association and its growing activities and to give him names and addresses of bird lovers in widely scattered locations who would make good members. The roster of the first hundred members of the ABA read almost like a who's who of the newly recognized birding world.

For Red and me for more than a dozen years it was a mighty fine journey, and along the way some good things happened. Red and I gave programs from here to there and back again. Using Bird World as their meeting place, I helped bird people meet other bird people to form the North Texas Bird and Wildlife Club in Wichita Falls. Using the same Bird World technique, I helped Victoria birders organize the Golden Crescent Nature Club. Both organizations are still functioning today.

I taught a summer enrichment class about birds and nature for young disadvantaged kids in first to fifth grades. Red helped me to take busloads of them on field trips to the sewage evaporation lagoons to discover birds with binoculars and scopes.

Newspaper photos helped promote the first Beginner's Bird Walk in Austin, Texas, organized and led by Marjorie. Participants in the walk ranged in age from eight to eighty. Left to right: Matthew Carson, Marjorie, Joe Wilson, Fritz Stanley, Julie Valentine, and Red Adams. (Courtesy of Austin History Center)

I also started Austin's first Beginner's Bird Walk (with participants aged eight to eighty) with the guarantee of ten bird species. We got eleven. And I taught fully booked classes on bird identification at Austin Community College.

With advice from Kathleen and Bob Zinn, ham radio operators of Wichita Falls, I tried to organize ham radio operators into at least a Texas-size rare bird alert. It is part of a license permit requirement that a ham do some public service, so it seemed a good match. Geth White of El Paso used a ham radio to report a Costa's Hummingbird seen and photographed in a nearby canyon. Doris Wyman of Port Lavaca reported more than a thousand Broad-winged Hawks flying over Woodsboro and Mountain Plovers in a rice field at Port Lavaca. Kathleen Zinn reported a Northern Shrike in Wichita Falls. I sent a message from Texas to California, and there were a number of other successes. However, the major problem was that the radio ham operator had to pass the message on by phone to a designated birder in his area—too many steps to deal with unless there was a great deal more organizing. Now e-mail can handle it in minutes.

Along the way, my column Bird World itself began to evolve. Longtime

Marjorie's classes on bird identification attracted capacity enrollment at Austin Community College. (Courtesy of Austin History Center)

birders will understand easily what I mean, because most have experienced the same evolution from being hunters and watchers of birds to becoming conservationists and, finally, activists. Seeing birds and other life struggling to make a living in the wild world, we come to realize fully that we are part of the process and that habitat preservation is vital to all of us.

Bird World consistently began to present the visions and goals of the con-

Guardians of the List. Left: *Edgar Kincaid Jr., editor of the noted two-volume* Birds of Texas, *disguised here as a Double-wattled Cassowary (courtesy of Cushing Memorial Library and Archives, Texas A&M University).* Right: *Fred Webster, for thirty years the South Texas editor for* American Birds.

servationist. This viewpoint was hazardous in those times, and one editor candidly admitted that in his area it was a viewpoint he could not pursue and thus he could never buy my column.

I remember feeling I was risking the loss of all of my papers when I submitted a column explaining (in my allotted page and a half) the Texas Water

Plan. This boondoggle was supposed to cure dry West Texas by piping water there from wet East Texas. To my relief, every one of my papers published it, and one editor went so far as to tell me he hadn't fully understood the plan until he read Bird World.

The times were such that Bird World was allowed less than two typewritten pages to inform, excite, dramatize, or proselytize. Its publication always teetered on the brink of extinction, but I think I kept on writing it under difficulties and restrictions because I just couldn't help myself. It was a sort of diary tool that I used to educate myself along with my readers.

Unfortunately, this education started with a blow between the eyes on the very first day Bird World met the public. The Austin paper gave me a wonderful send-off, with a notice on the front page that Bird World was beginning that day. Included were some facts about Marjorie Adams' qualifications, and it wasn't too long before I got a phone call from my brother, Jim Valentine.

"Marj, are you really a Fellow of the American Ornithologists' Union?"

I called Edgar Kincaid Jr. None other than the great Roger Tory Peterson himself noted that Edgar Kincaid was "the most thorough fine-toothed comb among my critics . . . as he gave [the Texas field guide] manuscript a thorough going-over." In addition, Edgar's wide acquaintance with observers around the state enabled him to advise Peterson which records of bird sightings could be trusted and which were "unsanitary." Edgar was not only a brilliant ornithologist, but he had a rare wit and had formed the habit of giving fellow birders nicknames, his own being "Double-wattled Cassowary." So here I was (the humble Chatter Duck-no-such-bird) on the phone with this hero.

"I think I've committed a crime, Edgar." Then I had a difficult time continuing.

"Yes?" Cassowary prompted.

An old memory was racing through my head. It was Cassowary's uncanny sense of drama, which could come forth at sacred events such as Christmas Bird Counts. This combination of character thrust on him the responsibility and indeed the duty to be sure that all reports were accurate or, as he put it, "sanitary." His questioning could be so protracted, detailed, and repetitive that the "guilty" observer had to finally come to terms with exactly what he actually had seen. These "sanitizing" dramas, although civil, were likened by some participants to procedures on the criminal witness stand.

Alongside Edgar sat Fred Webster, for thirty years the South Texas editor for the publication now titled *American Birds*. If Edgar was the judge, Fred was certainly the jury in cleansing, clarifying, verifying, and recording bird sightings.

Thus it happened that early in my birding life we were shorthanded on a Christmas Bird Count, so I undertook a territory alone. My sanitizing began when I reported a Green-tailed Towhee.

"*You* saw a Green-tailed Towhee?" Cassowary asked, looking me as straight in the eye as a fishing heron. "You know it's a rare bird here, don't you?"

"Well, I don't know how rare it is," I confessed guiltily, "but I saw *two* of 'em."

Perhaps it was only an hour later, but it seemed eons before my two towhees were finally accepted. I still look back on that day as a triumph.

And now I was confessing a real crime to the Great Cassowary.

"Edgar, what does it mean to be a Fellow of the American Ornithologists' Union?" I blurted out.

There was a long silence as I hung there in space, my apprehension doubling with each second.

"A Fellow at the AOU?" The pause continued to a torturous length. Then he helpfully answered, "Why, Marjorie, that's merely someone just below the level of Jesus Christ."

"Oh," I gasped. "Oh-h-h-h-h-h," I moaned. "I thought it was just another word for member."

I had disgraced myself. I could even be called dishonest. There was only one option: I would have to jump off the granite state capitol of Texas, as hari-kari was too bloody.

I sweated out a carefully worded explanation and apology, which I mailed not only to my editors but also to dozens of knowledgeable bird folks.

That was the not-so-noteworthy beginning of a little primer used to teach myself, and my readers along with me, how to discover an amazing world. To relive some of it, we'll start here with a facsimile of Bird World's first column as it was presented to the newspapers.

MY FIRST BIRD WORLD COLUMN

Why Watch Birds?

For release on or after November 1, 1965

WHY WATCH BIRDS?

What wild animal will you see today?

For most of us, the only warm-blooded truly wild creature we can be sure of almost anywhere, in any season or adversity, is a bird, and birds' easy availability and great variety are major reasons there are more birdwatchers than giraffe- or whale-watchers. More people observe birds than they do any other form of wildlife on earth, as evidenced by sales of bird publications, recordings, and binoculars; by capacity audiences at wildlife films and lectures; by the expanding rosters of nature camps and clubs; by bird-finding tours offered to nearly any point on earth; and by the many families who have discovered field trips. Nowadays, even business or retirement locations may be chosen under the influence of birds.

Birding has developed into a sport that, like fishing or football, offers its own excitements: the lure of discovery and the surprises and rigors of a hunt without the aftermath of death. There is no limit in pursuit of birds, except the planet itself, and few corners of the earth are dull. Yet birding can be hand-tailored entirely to the individual: it can be a dangerous adventure in jungles or on arctic cliffs, or an invalid can entertain feathered guests at a windowsill.

Still, there is more to it than sport. When a birdwatcher says, "It's like going to church," he is trying to describe the spiritual and aesthetic lift he gets from his communion with the outdoors. Tensions vanish as nature calms and inspires. Birds have such manifold appeal: their colors and designs can compare to great paintings, their flight and movement to the ballet, their melodies to concerts; they do many things we humans do, most of it engagingly; and they are a constant symbol of freedom, which gives deeper meaning to our seasons

Bird World

By

MARJORIE ADAMS

P. O. BOX 2124, AUSTIN, TEXAS 78767

and wonder to our spirits. In addition, a birder can serve through his hobby, for ornithology is one of the few fields in which the findings of the amateur often are of great value.

These are some of the obvious reasons for enjoying birds, but more profound rewards can come from their study. As man strives to understand birds (or other wild kinfolk) in their ceaseless struggle for existence, he soon perceives his own great power over living things. A larger philosophy is born when we fully realize that how we use this power may decide the fate not only of other life but of our own as well.

A question about birds? Send it to BIRD WORLD c/o this paper for a professional answer.

HOW TO PLAY THE BIRDS

Back in the 1960s at a busy meeting of the Texas Ornithological Society in San Angelo, an old gentleman sitting in the hotel lobby seemed to be neglected, so I strolled over and asked where he hailed from.

"Oh, I live here. I'm not a birdwatcher," he answered, and then he added with a grin, "I just came over to look at all the getups."

Remembering Red's various seasonal camouflage outfits and my own hooded trench coat, which I hand-painted with camouflage, I believe it's true: special activities do require special clothes, special equipment, and perhaps even special outlandishness, so it's fair to classify the old gent as a people-watcher.

It was not surprising that same day more than forty years ago that a lady getting on the elevator pointed to our binoculars and asked, "What do you *do* when you watch birds?"

I started to say, "Ma'am, that can't be answered on an elevator ride," but I smiled and replied, "It's a detective story—we try to track 'em down."

The proper answer, of course, was "That depends."

The salesclerk at Smith's Department Store who feeds pigeons on the roof during her lunch hour and the person who puts out a bird feeder in the backyard are both birdwatching. The scientist recording birdsongs deep in the Brazilian jungle is doing bird research. The person who is birding is playing a game and sport.

Like other sports, birding has rules and objectives, but when I began publishing Bird World in 1965, there was no central entity dictating rules. In fact, when Jim Tucker sent out his first organizational letters to birders to form the American Birdwatchers Association, he asked for help in putting together the rules for the game. Birders not only opted for the term "birding" instead of "birdwatching" but also came up with rules that have remained basically the same for years.

The game centers on the Life List, a lifelong record of all the bird species that one has correctly identified in the wild anywhere in the world. On the face of it, such a seemingly simple (eccentric? idiotic? useless?) objective seems scarcely worth mentioning, but strong men can succumb to this sport like a fever.

The challenges are multiple. They begin with hunting and finding birds in their thousands of varying habitats. The challenge continues with the birder narrowing down the hundreds of possible identification points, including seasonal and age changes in plumage, to reach the correct identity. The hunt is constantly compounded by a bird's ability to flit in and out of sight or disappear altogether. Add the beauties, complexities, difficulties, and dangers of the playing field known as the world, and the situation can build an emotional intensity that can change the habits and interests of a lifetime. Only half-facetiously, I date events in my life as B.B. and A.B.

Birding can be made into a variety of games by limiting the territory, by setting a time limit, or by setting different objectives. Roger Peterson told me he believed that thousands of adaptations are possible. For the Life List the world is the territory, but territory can be restricted to a country, a state, a county, a wildlife refuge, or a backyard. A bird must be identified only within that chosen territory to qualify for that particular list. A handicapped person can play from a window; the dauntless hero can climb mountains or tread dangerous jungles. Birding is one of the most adaptable games and sports in the world.

It is also one of the most democratic, embracing people of all ages from all walks of life. Even more noteworthy, it has changed much of the wild world into a democracy too: a little brown Grasshopper Sparrow gives a player exactly the same score of one as a gorgeous Resplendent Quetzal does.

Birding is always played on the honor system, and in the early days as the sport was being fully defined, this was the area that was most troubling to its organizers. Over and over the question was asked, "How do you know a player isn't cheating?"

It was common for us to worry, "What can we do to keep players who don't know the difference between a buzzard and a bullfrog from messing up the game before it is firmly established?"

It is logical here for me to turn to the founder of the American Birding Association himself, James A. Tucker, whom I dubbed "the father of us all" at our first ABA convention in 1976.

"Marj, as a charter member you know how many times all of us have been asked, 'How do you know a player is honest?'" Jim responded not long ago. "I've been asked just as many times or even more, 'Why and how did you start the ABA?'

"The ABA was the end result of a natural chain of events," Jim continued, "one thing leading to another. Back in the 1920s in a small college in southwest Michigan, the dean of women, Mary Lampson, was a bird lover, and she introduced several students, including my mother, to the hobby of birdwatching. It happened that my father loved the outdoors too, so he and my mother often wandered the woods to enjoy nature and search for birds not only in Michigan but in ensuing years also in Indiana and Illinois. There's no doubt I went on my first bird walk in my mother's womb, so you might say I was born a birder, and there have always been birds in my life."

"I wish I could have started that early," I responded with envy.

"Well, you're making up for it," he said charitably. "My earliest memory is of a White-breasted Nuthatch clinging to the trunk of the apple tree outside our kitchen window. I was three years old. My next birding memory is of a mockingbird. I proudly plucked its nest from a small shrub and took it to my mother. She promptly took the strap to my behind and put the nest back. That's how I learned that birds were to be protected.

"I can remember looking at a bird and wondering what was going on in its mind. I was wishing I could talk to it.

"When I was eight, the same Mary Lampson who had introduced my mother to birds retired and moved into our neighborhood in Nashville. Now she turned her missionary zeal on me. She told me how she and her friends had always tried to list a hundred different species by May 1 of each year. That became my goal too. At nine years old I was ready for listing."

"Did you actually keep a written list?" I asked.

"Yes, I did. As time passed I became a member of the Tennessee Ornithological Society, an active bird bander, a museum curator for a small college, and a teacher of bird classes at a camp. Listing had become a passion that brought me in contact with other birdwatchers, and while I was in college, I became president of the Chattanooga Chapter of the Tennessee Ornithological Society. I was twenty-one years old. When I started teaching biology, I chose Florida because of its rich birdlife."

I noted, "You were setting that pattern early in birding history. I'm sure you know that nowadays it isn't that unusual for a person to choose the headquarters of a business or a retirement locality for exactly the same reason."

"It was a good move for me," Jim observed. "I became local editor there for *Audubon Field Notes* and was elected president of the Orange County Audubon Society. Now I was involved with both birds and conservation."

"That's an almost universal progression now," I said, "and I think it's the most constructive aspect of the game."

"Yes, I've seen it happen a number of times. It was part of my job as president to introduce the speakers who brought the Audubon Screen Tours to our town," Jim reminisced, "and one of them was G. Stuart Keith. He had written an article in *Audubon Magazine* about birders who had identified more than six hundred birds in the field in North America. This had set a goal for birdwatchers, and it was a thrill for me to meet him. I gave him a copy of my newly published *Combination List and Checklist for Birds of North America*. Stuart praised it because in notebook form it lets you record not only your Life List but also annual and state lists and even migration dates all in a compact form."

"That unique little book is also what brought you and Red and me together," I reminded Jim. "I reviewed it in Bird World, and we also showed it in our movies at an Audubon program. That's when you happened to be in the audience and came up afterwards to introduce yourself as the author. But go on with your story."

"The upshot was that Stuart and I made a pact that night in Florida to race to the highest score in North America that year. I don't remember who won—I think he did, but I certainly remember we were kindred spirits.

"About that time," Jim recalled, "I was awarded a National Institute of Mental Health predoctoral fellowship in school psychology at the University of Texas in Austin. So there I was in the summer of 1968, separated from the friends who had become such a large part of my life in Florida.

"I guess I was lonely and I needed a way to communicate with them or other bird chasers. I realized some birdwatchers were conservation types and others were scientific types. Neither pastime seemed to be connected with the same kind of fun that sport types were having. Why couldn't it all be put together?

"Maybe it was the listing race with Stuart Keith, or maybe the growing competition in Christmas Bird Counts, but it seemed to me that chasing birds could certainly be a game or even a sport. The result was the exploratory newsletter I sent out in December 1969 to bird people I knew, including you and Red, asking them to become members of the American Birdwatchers Association at three dollars each."

"It was purple," I said with a laugh. "You duplicated it with that gel-and-ink system that teachers used in school. I still have it, a little piece of history I'm hanging on to."

"Well, it was a start," Jim said humbly. "The basis of the game was listing in specific territories. Stuart was immediately supportive and suggested that the name be changed to American Birding Association. The genius of his insight was that the term "birdwatcher" belonged to an image of the past that we wanted to avoid. He also sent me a long list of birding friends and urged that I include them in the mailing list. This led to Arnold Small, Bob Smart, and Joe Taylor becoming involved. So there was the embryo, and with the help of other bird aficionados the basic game took form over a period of months.

"When we finally got to organizing officially, I wrote to both Stuart Keith and Arnold Small, suggesting that one of them be president and the other vice president. They agreed that Stuart would be the first president and Arnold the VP. We asked Joe Taylor, who was approaching an extraordinary North American score of seven hundred species, to be treasurer, and Benton Basham would be membership chairman. I would serve as secretary and newsletter editor. Before long we were thinking about a national convention. The rest is history."

"So I guess, Jim, you were an inventor," I noted. "I'm thankful we happened to be there at the right place and time to meet you and join in the fun. It's one of our biggest rewards."

It truly was a special time for many of us, and I'm remembering back to the Jim Tucker of those days. He was thirty years old but looked twenty. He was slender but athletic, ingenuous, and creative. He was living with the problem of a son suffering from cat scratch fever, which progressed into encephalitis (and, in years, to invalidism), and this personal problem was directing Jim's life career to special education.

Deeply religious and perhaps needing it himself, he was creating a book called *Windows on God's World,* a daily devotional with a selection from Scripture accompanied by an adventure or observation of God's natural world that was attuned to it. The book was published in 1976, a leap year, so it included 366 devotionals. Since its publication, more than a hundred thousand copies of this book have been sold on five continents in five languages.

Through the years, Jim has continued this side of his life with other religious writings as he worked with education for the handicapped. Presently he fills an endowed chair in learning exceptionalities at the University of Tennessee at Chattanooga. And he has never stopped his joyful pursuit and study of birds in the natural world.

Let's go back here to the question, How do we know a person is playing the game and sport of birding on the honor system?

"Jim, we all knew from the beginning this could be a really sticky problem," I said.

"Yes," Jim agreed. "And it was a big step forward when Keith Arnold, professor at Texas A&M University, founded the Texas Bird Records Committee in 1972. This gave definite rules for confirmation of sightings. I learned a lot from serving on the committee myself. We exercised the strictest rules on identification.

"At ABA everyone of us had previous experience with people with various degrees of field ability," Jim continued. "We all knew that bird identification is a skill that must be learned. So one way to curtail poor identification by novices was to keep membership limited to skilled and qualified players. So we set up membership categories. To become an active member, you had to have an AOU Life List of at least five hundred species or a state or province list of 70 percent or more of the species on the list for that area. The one hundred charter members of ABA all qualified in this way. We also had associate memberships for those with lists of four hundred species or 60 percent of the species on a state or province list. However, we soon dropped this category.

"By the time of our first convention, we had set up a governing body called elective members. Marjorie, you from Texas and Olga Clarke of California were the only women in this group. Only elective members could vote on organizational matters, including the qualifying of new members. A person needed to have this approval before he or she could be elected into membership."

"I remember very well our open discussions about proposed new members," I said. "We needed and wanted members, but we leaned over backwards to screen new prospects even though most prospects had already been screened to some extent before their names were submitted. We were anxious to grow, but fearful we'd get off to a bad start."

"Well, the system worked—and we grew," Jim recalled with satisfaction.

Nowadays, with the sport of birding firmly established, all it takes to become an ABA member is a written request and the submission of annual dues.

Reports of bird rarities have always required careful field notes, voice recordings, or other substantiation. In an interview I had with Jim in 1973, he said, "If we were solely competitive in playing the game, we wouldn't tell another soul about a rare bird we found. But when we tell our competitor about it, he's going to verify the identity of our own score. That's certainly a built-in feature for keeping birding honest." Today so many birders may show up for a rarity with the hope of adding it to their own lists that there is no lack of witnesses.

Through the years, birding rules continued to be debated, clarified, and refined. One of the first questions was whether we would have our own ABA species list or whether we would be governed by the species list scientifically defined by the American Ornithologists' Union. The decision was that though we were not a scientific organization, it would be most beneficial both to the birder and to science to abide by the scientists' decision as to what constitutes a species.

No one knew, of course, the advances that would be made over time by DNA testing and other methods of defining bird relationships. We have groaned as tests lumped together birds we thought were separate species, and then we have gloated when tests split what had been considered a single species into two or even three. The categorization of species is a continuing challenge.

By 1971 a question arose that was causing dispute not only in the ABA Rules Committee but within the general ABA membership: Can you count a bird that you have not seen and have identified only by its vocalization?

Since most of us start our birding by sight alone, the debate came as a surprise to many, yet logical arguments for counting heard-only birds were being advanced by highly qualified birders. Some of the arguments, pro and con, were published in *Birding* magazine.

Edwin I. Stearns of Westfield, New Jersey, noted, "It is possible for a blind man to have a Life List and to enjoy birds. The real test as I see it is whether or not the field mark is characteristic—be it sight or sound."

Dean Fisher of Nacogdoches, Texas, declared, "It requires just as much skill, if not more, to identify birds by their songs or other vocalizations as it does to learn them by their outward appearance. I am not concerned with which of my senses I use to make the correct identification."

Noel Pettingill of Houston, Texas, stated he had seen all the birds on his Life List, "but sound is essential and therefore equal in importance to sight in identifying certain species."

Shum Suffel of Pasadena, California, asserted: "Positive identification is the criterion for placing a bird on any list, and since in many cases this is best accomplished by call notes or songs, then sound alone can constitute identification."

An anecdote by Roger Peterson made it seem reasonable to count heard-only birds. Peterson said he lay in bed one morning while the rest of his group got out at dawn searching for birds. "Lying abed," he reported, "I identified by song more birds than my early-rising friends identified by sight."

For me, seeing was what counts, and playing by this rule caused me a ridiculous amount of effort and painstaking research for a bird that by sound-only was already a cinch. Lying in bed in Austin, Texas, where the Whip-poor-will is a rare migrant, I had heard the bird's unmistakable calls for the first time. I could have counted it as a lifer without even rolling over, but it was not until several years later that luck gave me my first sighting of a bird that was a possible Whip-poor-will.

This sighting was on the Pedernales River, thirty miles west of Austin, where a Whip-poor-will would be even rarer. At the time, Red and I were writing and filming and living in the "Road Roost" (our twenty-two-foot travel trailer), perched on a bluff above a canyon dug through the ages by Roy Creek. At its confluence with the Pedernales, the canyon had a depth of two hundred feet or more, and this "difficult" bird was near the bottom of the chasm.

After looking the bird over meticulously, I climbed the steep trail up the bluff to my books. I easily eliminated the Common Nighthawk, but was the bird a Whip-poor-will or a Chuck-will's-widow?

With a flashlight shining on it, I have watched a Chuck-will as it was calling, but that doesn't give much detailed information for a daytime view. All the guides mentioned the similarities of the two species, both being much mottled and heavily vermiculated but with one more brownish and the other more grayish—not much help in mixed sun and shade with a bird sleeping not in the horizontal as pictured in the books but in a somewhat upright position and firmly settled in a tree against a huge boulder that blocked access to its back.

As for size, there wasn't much to compare it with. So back up the trail I went, to study Bent *(Life Histories of Familiar North American Birds)* and the recently published Oberholser *(The Bird Life of Texas)* and to make a phone call or two.

Again I slipped down the rough trail into the canyon where the bird slept immobile. I stared at it quite some time. "Why don't you say something?" I whispered.

If I roused it, it would it be gone. I wished for Edgar Kincaid, Fred Webster, and John and Rose Ann Rowlett, then just sweated it out, finally realizing I had a female Whip-poor-will, definitely grayish brown and about the size of my nine-inch-long tennis shoe. I had spent most of a day and had made three trips up and down that tricky trail in order to get a live body to go along with the call I had heard five years previous.

Arguments for counting lifers only by sight were also included in *Birding*. P. William Smith of Matawan, New Jersey, wrote: "I don't know any bird *that* hard to see, especially with calling tapes nowadays. There is always the chance of error."

Ira Joel Abramson, Miami, Florida: "Even the best descriptions of a bird's song leave a great deal to the imagination. And in some areas birds are excellent mimics." (For example, Red and I had European Starlings imitating Northern Bobwhites in the middle of downtown Miami.)

Charles T. Clark, Des Plaines, Illinois: "I do not feel they should be counted, mainly from a sporting challenge viewpoint."

Stuart Keith, Kenya, Africa: "For me, it is not a question of logic but of feeling. I actually want the bird in front of me before I count it."

For me, Arnold Small of Beverly Hills, California, was the most convincing. He wrote: "First, there is popular opinion and usage. Birding is essentially a visual game and always has been. We are called birdwatchers, not bird-listeners, and we use binoculars and not hearing aids. . . . And what about birds in foreign countries where calls are not so well known? Are we going to have a different standard or set of rules for each country? A bird noise (however pleasing) is *not* a bird; it is a disembodied sound. . . . A song as a field mark is not a bird any more than two white wing bars (if they could be disembodied) are a bird."

For me, Arnold's argument is even stronger if it is stated in this way: A bird vocalization is not a bird but something produced by a bird.

So if an unmistakable sound produced by a bird can be counted as the bird itself and thus as a lifer, what is to keep us from counting as a lifer a feather that we see fall out of a tree and is proved to have been grown by a certain species?

Think of the possibilities! The unmistakable nest! The unmistakable egg! The unmistakable footprint! Etc., etc., etc.

Time has proved the wisdom of the conciseness of the ABA Listing Rules. With some refinements added from the original rules (italicized in the passage below), they are as follows, as of the year 2005:

1. The bird must have been within the prescribed area *and time period* when encountered.
2. The bird must have been a species currently accepted by the ABA Checklist Committee for lists within its area, *by the A.O.U. Checklist for lists outside the ABA area and within the A.O.U. area, and by Clements for all other areas.*
3. The bird must have been alive, wild, and unrestrained when encountered.
4. Diagnostic field marks for the bird, sufficient to identify to species, must have been seen and/or heard and/or documented by the recorder at the time of the encounter.
5. *The bird must have been encountered under conditions that conform to the ABA Code of Ethics.*

The American Birding Association Principles of Birding Ethics is a double-column, detailed, full page addressed to both individuals and groups. It strongly states that, in any conflict of interest between birds and birders, the welfare of the birds and their environment comes first. A copy can be obtained from the American Birding Association, P.O. Box 6599, Colorado Springs, Colorado 80934-6599, or from www.americanbirding.org.

For the official North American list, birds must be identified in the territory recognized by the American Ornithologists' Union. This includes the forty-nine continental states of the United States, Canada, the French islands of Saint-Pierre and Miquelon, and adjacent waters to a distance of two hundred miles from land or half the distance to a neighboring country, whichever is less. Excluded by these boundaries are Bermuda, the Bahamas, Hawaii, and Greenland.

It happened that while I was working on this story with Jim, I found in my files the rough draft of a letter addressed to a Mr. Green. Don't know if I mailed it. It concerned a recent announcement in the Texas Ornithological Society's newsletter of June 1973, in which Mr. Green solicited contributions (in lieu of dues) from persons who were qualified and who wished to become members of the 600 Club. I wrote:

Mr. Green,

I am concerned about this development, and I feel many other birders across the nation will feel the same concern. Ever since 1963 when Stuart Keith published in *Audubon Magazine* a list of nineteen birders who had an AOU North American Life List of 600 or more species, I have felt this was the beginning of the "600 Club," as we birders have known it since that date.

The 600 Club immediately gave North American birders a definite and tangible goal, and I know birders who have since qualified by properly identifying 600 species in North America north of Mexico, and who felt that by doing so they had automatically become a member of the 600 Club, a symbol of achievement which has belonged to all the birders of America since 1963.

I find myself perturbed by the idea of an individual taking charge of the 600 Club and soliciting monies in the name of that individual for membership.

Though we have not met personally, I have been told by persons who know you that you are an individual of high personal integrity who is highly respected in his profession and who has made significant contributions to science. Had I not been so informed, I could easily misconstrue your motive in soliciting funds. I feel that when this is pointed out, you will not want to continue this solicitation.

If we ever do have such a thing as an official dues-paying membership in *the* 600 Club, it should by all means be administered by a recognized club or organization.

Mr. Green made no further solicitations.

Jim Tucker clarified the situation further: "During a spring meeting of ABA board members in Rockport, Texas, we met Earle Greene, and we all enjoyed our acquaintance. He was a trained ornithologist who had contributed much to that discipline, but he had retired by the time we got sport birding off the ground. He was delighted with things relative to the American Birding Association, but unlike Irby Davis, whom you knew well, he believed the whole sport should be based within the science of ornithology and not depart from the rigorous scientific scrutiny of that discipline. He played a significant role in being the first to try to build a 600 Club."

The question now is that with so many birders reaching lists of six hundred birds, does the 600 Club mean much anymore? Has it been displaced by the "700 club"? No, not for me, or for many others.

World birders have many decisions to make, among them which checklist of the birds of the world to play by. Yes, there are several (conflicting) lists, and more are coming. In a review in *Winging It* in April 1999, Charles Sibley wrote, "Choosing a checklist is like choosing a religion . . . the whole issue of species and subspecies has become a minefield."

At present the ABA is choosing the taxonomy of Clements. Specific and accurate observations become very important as sightings by capable lay birders add to the world's scientific knowledge of birds, so agreement on which checklist to use is important.

In an exchange with Jim Tucker, I asked this question: "As founder of the American Birding Association, you could have made yourself the association's first president. Why didn't you?"

"Such a thing never entered my head," Jim replied. "I don't enjoy administration and management. I like leadership, and you don't have to have a position to provide that on many fronts. For a long time our growing ABA was run by volunteers. The fact that I eventually became executive director was more by default than anything else. I was already doing the tasks of such a position, and the board made it official by making the appointment."

"But you were also taking the responsibility of the newsletter and other publications," I continued. "What was in it for you with this considerable workload?"

"I do like to write, and I like to communicate ideas and report achievements. Being editor of the ABA newsletter was perfect for me because it gave

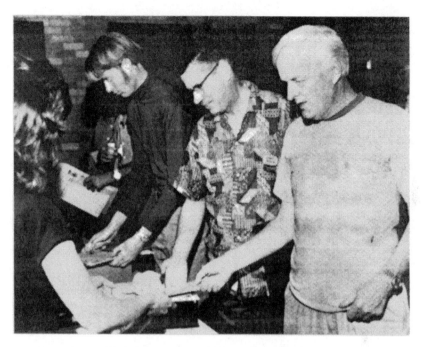

Left to right: *Bradford Blodget, along with Chandler Robbins and Roger Tory Peterson, authors of competing bird guides, as they register for the first American Birding Association convention, Kenmare, North Dakota, 1973. (Courtesy of* Kenmare, ND, News)

me what I wanted in the first place—ongoing and regular contact with my birding friends."

I found in my files these notes I made at our first ABA convention, held June 14–17, 1973, in Kenmare, North Dakota (population: 3,500). Listen to this and sigh: registration, $3; banquet, $6; and field trips, $5.50. Compare that to a typical ABA convention now: a full package cost of around $470, plus hotel expenses anywhere from $50 to $90 per night.

I also asked Jim this question: "Did you ever visualize that ABA would be so financially successful? You could have *owned* it!"

"No, of course not. I don't think anybody would have conceived it, and I don't believe anyone thinks of it as such now. ABA just grew along with the membership growth and with the ideas of all the people who made ABA a success. Unfortunately, as the business side grew, it demanded more and more of my time while I had less and less time to devote to it. The board was slow to recognize how important the business side had become. In fact, at one time the board wanted to close down ABA Sales, on the premise that sales

Nominated by Marjorie, James A. Tucker (left) was voted "Citizen Sportsman of the Year" by the Texas Outdoor Writers Association. TOWA president, Charly McTee, stands by as Marjorie presents the award. (Courtesy of Texas Outdoor Writers Association)

did not seem in harmony with the objectives of the organization. I believe that now there is general agreement that ABA Sales continues to offer many benefits to members."

"Here is something that is a little puzzling to me," I told Jim. "I think it was Paul Green who wrote in *Winging It* that the ABA was established for the traveling birder. Was this like anything you had in mind at the beginning?"

"Absolutely," Jim responded. "The ABA has been from the beginning an aid to the birder who wants to expand his or her experience with birds in a geographic way. However, you'll recall that my primary motive was friendship and companionship.

"The lure of the list, of course, demands at least some travel and movement (except for the handicapped confined to a window), but more fundamentally it is hard to understand the scope of birding without experiencing as many different habitats as possible. Experiencing new vistas and new sounds in search-and-discover missions forms the core of birding.

"And now there are a number of championship competitions around the country, such as the Great Texas Birding Classic, which covers the entire Texas

James A. Tucker, founder of the
American Birding Association.

Arnold Small, first vice president of the
American Birding Association.

Officers elected at the first convention of the American Birding Association. Left to right:
Roland Wauer, publications chairman; Benton Basham, membership chairman; Joseph
Taylor, treasurer; Stuart Keith, president; Jim Tucker, secretary; Ann Gammell, hostess for
the first ABA convention; Bob Smart, second vice president; and Olga Clarke, member of
board of directors. Not pictured: Chandler Robbins, Checklist Committee chairman;
Arnold Small, first vice president; and Marjorie Adams, member of board of directors.

ABA buses unload for a field trip to search for Le Conte's and other sparrows at Lostwood National Wildlife Refuge, North Dakota. (Courtesy of Bob Danley)

Gulf coast. And of course there's the competition in New Jersey. At last birding is recognized as a true, well-organized competitive sport."

"What would be your plans and dreams for ABA if you had the power to set its course?" I asked.

Jim thought for a long moment before replying, "That's an interesting question." He looked directly at me as he said, "I did set its course."

He paused again before he continued: "I would want ABA to be what the birders it serves want it to be. I am intensely pleased with the current course. Birders have collaborated to make it a resounding success story. I was a willing participant in the right place at the right time. I simply thank God for that chance and the thrills the experience has brought."

Robert and Ann Gammell of Kenmare, North Dakota, served as chairpersons for the first convention of the American Birding Association in 1973. However, the town as a whole became host to the conventioneers. Shop windows on the main street displayed handmade posters. Homes opened to take in the overflow from the two small hotels. The modern school building served as the central meeting place, and the school gymnasium and showers were available to birders who brought their own bedrolls. Bountiful meals were cooked and served in the school cafeteria by the local matrons.

Everywhere, birders were greeted with smiles, and not solely out of hospitality, because the birders spent many good dollars in Kenmare. The town had found a source of riches (and not just money) that it had never exploited.

Since 1957, Bob and Ann Gammell had banded thousands of birds, and bird disciples from all over the United States and from foreign countries had brought a part of the outside world to their small-town door. When seven buses full of bird questers unloaded in front of the Gammells' home, their neighbors came out on porches to people-watch unabashedly. I chose that moment to ask Ann, "What have birds and birding meant to you?"

As she looked around at the large flock assembled from all around the country and standing there giving her homage, her eyes grew misty.

"I can't answer that question," she said, overcome with emotion.

BIRDING — A DANGEROUS SPORT?

Stuart Keith was unquestionably the champion birder of the world back in the 1970s. He had stalked birds for twenty-six years in forty countries and on every continent to tally more than five thousand species. What he told me then is still applicable today.

"Worldwide birding involves all kinds of challenges," he said. "A player must study beforehand the birds he hopes to find in foreign lands, but if books or references aren't available, he must glean his information from museum specimens or from an occasional local naturalist. Alien customs, languages, and immigration officials must be dealt with.

"And I've used everything from a camel to a swamp buggy for transportation. Sometimes a guide may turn out undependable, and often information isn't accurate. You have to take precautions against disease, dangerous animals, insects, and snakes. Even a minor injury far from medical attention can mean hardship or tragedy."

Keith's record is proof that birding is no sissy sport, but you don't have to go abroad to prove it. Following are some other adventures from my file on dangers.

"As part of my birding life I do bird banding," Red Mason of Etobicoke (a suburb of Toronto, Canada) told me back in the 1970s. "I had managed a hard climb up an eighty-foot tree to the nest of a Great Horned Owl for the purpose of banding the young owlets. I made sure not to look down as I took a breathing spell to figure the best way to get hold of an unfriendly youngster. Then as I reached for its leg, BAM! was my greeting from the mama owl as she returned to the nest. She whirled and launched another attack, knocking off my safety glasses.

"When I tried to catch the glasses, I swung off balance, and before I could get straight again, BAM! The big owl came at me with those terrible claws. This time she took my hard hat.

"Without protection for eyes and ears I decided it was not very smart to challenge one of North America's biggest birds of prey. I gave up, but she didn't, and she flew at me again and then again as I worked my way down. That tree must have been one of the fastest-growing in all history, for I felt sure it was at least five hundred feet tall before I finally got my feet on the ground. I was bleeding in a number of places."

* * *

I was sitting in the office of Pronatura's headquarters in San Cristóbal de las Casas, Chiapas, Mexico, when John Sterling, who was doing bird censusing for the Smithsonian Institution, walked in, sunburned and weary.

On a large wall map of Mexico he traced his work of the last four days. He and Peter Bichier Garrido of Argentina had searched the countryside almost to the Guatemalan border and back up along the Pacific coastal plain, a trip of about 350 miles. John was pleased that on this census they had seen the endemic Giant Wren.

"John, do you feel safe camping out here?" I asked. "I know you don't speak much Spanish."

"Indian would probably be more help around here," John said with a shrug. "Yes, I feel safe.

"I guess the most serious danger I've experienced was driving on an icy road. My car rolled five times. But some folks helped me get it right side up again, and it would still run, so I went right on birding that day in Arizona, and I got my six-hundredth bird—a Buff-collared Nightjar.

"Birding in mountains is dangerous, especially in places like the Andes and California because of the steep terrain, rock scree, and slides. And in the Amazon *everything* is potentially dangerous, so you learn to live with caution but not fear."

In only a few months a real and desperate danger would face John and all of the other Smithsonian scientists working on the census. They were continuing their work from a house in the town of Ocosingo when on January 1, 1994, the Zapatista National Liberation Army made simultaneous attacks on San Cristóbal de Las Casas, Las Margaritas, Altamirano, and Ocosingo. This life-threatening episode is recounted elsewhere in chapter 31.

* * *

Birders, unlike hunters, are not armed. Years ago, Bruce Broadbrooks of Los Angeles told me he was birding on the Texas coast at Smith Point, a location famous for migration fallouts. It was rainy and really soggy, and suddenly he was looking into a double-barreled shotgun held at a comfortable shooting level.

"I can shoot you if I want to" was the gunner's greeting.

"That's all the landowner said," Bruce told me, "and that's all he needed to say as he kept pointing the gun at me. He followed right behind me every step until I marched to the road."

Just to make Bruce feel better, I think I told him it was lucky this happened in East Texas and not West Texas. But perhaps I should have also added South Texas and possibly North Texas. Texans have a thing about guns and, likewise, a thing about trespassers.

***

Champion birder Joe Taylor had many close calls. He persuaded a bush pilot in Alaska to fly him to the tundra. When it came time to land, Joe discovered that there wasn't a lake long enough for a plane with pontoons to land. That didn't seem a problem to the pilot. His solution was to bounce the plane on land before hitting the water.

On one occasion the ship Joe was on managed to break free just as an ice pack was gripping it off the coast of Greenland. In pursuit of the elusive White-tailed Sea Eagle, Joe got lost in a remote part of this same Danish island. In 1956 he was washed off a boat while trying to land on the Santa Barbara Islands, wintering home of the Rhinoceros Auklet.

***

Sometimes a spur-of-the-moment decision can lead to a dangerous situation. Richard Albert, a surgeon in Alice, Texas, had suggested to Kent Rylander of Lubbock that they hop into Richard's small plane for a quick flight to the Gulf coast. The hope was they might catch sight of the then very rare and endangered Peregrine Falcons as they migrated down the coastline.

Kent had already doubted his decision when his pilot made a shaky landing on a windy and isolated beach south of Freeport. Now he was greatly regretting he had accepted this invitation. Richard had pointed to his leg and calmly announced, "Rattlesnake. Under that little bush over there. Got me right here on my leg."

When the two birders had left a meeting of the Texas Ornithological Society in College Station, their stated goal was the rare peregrine. However, Richard had soon decided to bird in the sand dunes instead.

"Although death from the snakebite didn't seem an immediate problem, I was sure wondering how we were going to get out of this mess," Kent recently told me, many years after this adventure. "I felt sure Richard would soon be so incapacitated by the snake's venom that he couldn't fly his plane. I envisioned being stuck a long time in this remote spot with a very sick person.

"But Richard didn't seem worried at all and wanted to keep on birding. I told him a flat no on that, and we headed back to the plane, which was as much as a mile away. Soon Richard stopped and said he wanted to go back to find the snake, kill it, and have me take his picture holding the snake next to his leg. By this time he had to slit his trousers to accommodate the leg's swelling, and his gait had become much more labored.

"Richard didn't find the snake but had me take a picture of his leg anyhow, which had two streams of blood trickling down it. Finally, we were again headed back to the plane, and all the while Richard was explaining the basics of flying to me. I had never been in a small plane before—in fact this was only my third time in any kind of plane—and I tried to inscribe on my brain every single word he said.

"By the time we got back to the plane, Richard was drowsy, but he showed me how to manipulate the controls and together we managed to take off.

"As we flew over Galveston Island, I suggested we land for an antivenom treatment, but Richard wouldn't consider the idea. Instead, he pointed out that we were getting low on fuel!

"He was getting more and more drowsy, but together we landed the plane at Texas City and got the tank filled up.

"At last we were headed for College Station, but Richard now was extremely drowsy. I kept watching to see if he was passing out and I kept asking him to repeat instructions on how to land the plane. It seemed a miracle, but somehow he pulled out of it enough that together we landed the plane—bumpily—but we made it.

"At the banquet meeting that night all the chatter focused on Richard's empty president's seat at the head table. Then suddenly, as the vice president was making announcements, in limped Richard, stone-faced as he seated himself, as though nothing had happened. One person told me later that she expected Richard to die at any moment in front of us all. Of course, he didn't.

"I learned later that Richard's ability to cope with the bite could have been due to a partial immunity he had deliberately acquired to rattlesnake venom. Reportedly, he had held a rattlesnake up to his thigh and let it bite him in order to build antibodies. How I wish I had known that as we were walking back to the plane that day!

"Next morning, though he hobbled, Richard was out in the field again. It was then that folks began to tell me about some of the doctor's more colorful exploits. One of them was that down in Mexico he had walked fifteen miles to get to a volcano that was erupting, and he had actually climbed partway up it as it was still boiling and flowing."

* * *

The late Arnold Small of Beverly Hills, California, shared this adventure with Red and me one afternoon. Arnold was on a bush-walking expedition after birds in northern Kenya when he and his guide discovered fresh rhinoceros tracks in a mud wallow. Suspecting that the fractious beast might still be near, they were moving with utmost caution.

Suddenly the angry trumpet of an elephant blasted forth not more than eight feet away.

"It was a nightmare," Arnold told us. "We'd been concentrating so completely on the rhino, we didn't see the elephant leaning against a tree asleep and just fading grayly into the landscape.

" 'Run like hell, Arnold!' my guide yelled. 'He's after *you!* Throw your hat down! Throw your hat down, *and don't look back!'*

"Without missing a step, I jerked off my hat and threw it high in the air. Miraculously, the giant animal stopped, picked it up, smelled it, then threw it down and stomped it. This gave us time to get away, and I can't remember whether we found any birds that day."

* * *

Harry Miller of Austin recalled a childhood adventure for us one afternoon. "I was about eleven years old," he began, "and on a dove-hunting trip with my dad and three of his friends out near Midland. We were having a wonderful hunt over an old maize field with a watermelon patch along an irrigation ditch.

"I was shooting a .410 and doing well. Every so often I'd quit hunting long enough to sit down and eat watermelon.

"The men decided they had their limits and it was time to head back to Midland. They were in a four-seater plane and thought I might like the plane ride, a quick way home.

"I thanked 'em, but I was having too much fun with hunting and watermelon, so the plane took off down the road without me. As we were still waving good-bye, the plane stalled and hit the telephone wires, flipped over, and landed upside down. All the passengers died instantly, with broken necks.

"That's the first time I saw a dead person. It hit me hard that I could have been one of them. It was then I realized that the Mourning Doves I had been shooting at all day had saved my life."

***

As I write this February 10, 1971, after yesterday morning's terrifying earthquake here in the Los Angeles area, there are frequent aftershocks to keep the realization fresh that we humans haven't achieved as much as we thought toward conquering nature.

Our services have not been needed in the disaster's rescue work, and since it is impossible to communicate with our son at his sure-to-be damaged home in the evacuation area, we decided to divert ourselves from our distressing anxieties and fears by attending the meeting of the Los Angeles Audubon Society, which, to our amazement, was still scheduled.

Young, handsome Charles Collins, professor of biology at Long Beach State College, had the opportunity to prove his excellence as lecturer and photographer on his specialty, the swifts, for midway through his presentation a hard aftershock shuddered the darkened building to its high, heavy beams.

There is no describing the agitation and fear one feels as the earth moves, but since the brilliant ornithologist calmly went on explaining his thesis of how several swift species managed to coexist on the island of Trinidad without competing, I merely moved to the edge of my chair, into racing position.

And race I did—halfway up the aisle, along with others—when an even stronger trembler jolted us.

But our gifted lecturer had held fast, and in fact the tone of his voice didn't vary one note as he continued his lecture without pause, and, ashamed of our cowardice, most of us returned and sat the lecture out.

However, I don't recall much of its end, for I was deep in a reverie concerning the utter devotion of bird lovers and conservationists on a day of earthquake, *Apollo 14,* and a moon eclipse.

***

Stuart Keith's foreign travel dangers didn't include the kind encountered by L. Irby Davis on one of his many trips to Mexico. Irby was a pioneer in recording birdsong, and his method was to camp primitively out in the wild. On one trip not only was he robbed of his equipment and money, but the thieves also took his car, leaving him stranded deep in the jungle.

Irby invited Red and me to go on one of his trips. My emotions said yes, yes, but common sense told me I didn't have the physical stamina for it, and Red wouldn't go without me.

The adventure of a lifetime, the road not taken. It does happen.

***

Certainly one of the most frightening adventures a birder could have took place March 23, 1998, about thirty-five miles south of Bogotá, the capital of Colombia.

In 1997 a group of birders in Colombia had achieved the remarkable feat of recording more than 1,030 bird species in one month in one country. Certainly that record is a testimonial that Colombia is the birdiest country in the world. (For example, Colombia has 160 species of hummingbirds.) This birding group's achievement was also intended to demonstrate that Colombia was now a safe place for birders.

Louise Augustine of Chillicothe, Illinois, and Todd Mark of Houston were members of this fortunate group, and they returned to Colombia for the third time in March 1998 with two other birders, Peter Shen and Tom Fiore, both of New York City.

On the last day of their trip the four birders were seeking the Cundinamarca Antpitta, a small robin-size ground-dwelling bird known to science only since 1989. Their search led to a dirt road named Antpitta Road by birders, and six miles later, as it was just getting daylight, they came upon a stalled truck. Soon they were surrounded by perhaps twenty men in uniforms and all carrying guns. They were members of the Revolutionary Armed Forces of Colombia (FARC), one of Colombia's largest guerrilla groups.

Thus began the kidnapping trauma of the four birders, which started with the loss of all their money, passports, binoculars, telescopes, walkie-talkies, tape recorders, and microphones. They didn't know it, but their presence had come near to interrupting a planned roadblock soon made by other members of the FARC on the main highway, an event that ended with twenty additional persons taken hostage and another four persons killed.

Nineteen of these hostages were brought to and held in the same area with the birders. One of these nineteen had hung onto a radio, and in a broadcast press conference the birders heard themselves denounced as suspected CIA spies who, if proved so, would be killed. In the meantime, it was announced, they were being held for a $5 million ransom. It is U.S. policy not to pay ransom or make concessions to terrorists, so the birders knew that government assistance would be limited.

In a conversation with me, one of the victims, Todd Mark, recalled: "The guerrillas had long-term camps with supply lines, and they moved us around. We were all holding up well, because world birding requires you to stay in good shape. Our diet was a lot of macaroni, a few beans, fried bread for breakfast, and sometimes potato soup. We had blankets or bedrolls for sleeping."

Remarkably, Tom Fiore managed to disappear and got an hour's head start before the guerrillas discovered he had escaped. To Fiore's surprise, in eight hours he met up with an equally surprised television crew from Bogotá. A few hours later, he was able to call the families of the other birders to assure them that the birders were alive and being treated well.

At age sixty-three, Louise Augustine, a retired teacher and former nun, was holding up OK until she fell from a rocky ridge and rolled several hundred feet down a steep slope into a gully. "It was far enough to kill her," Mark said.

She suffered eight cracked ribs, a fractured pelvis, and a collapsed lung. Though in severe pain, she was still able to walk with the help of poles cut for her by the guerrillas. At times she crawled.

The army seemed to be closing in on the guerrillas. "It was not a pleasant thirty-five days in a war zone with all those guns around," Mark told me.

One of the other hostages was Vito Candela, an Italian businessman who lived in Bogotá with his Colombian wife. He was released and promptly spread the news about the $5 million ransom for the birders and about Augustine's serious injuries.

The birders were now in what they called the concentration camp, where some hostages had been held six months or more. They rejoiced to learn that Tom Fiore was safe at home.

Augustine was taken by two FARC men out of the camp on the back of a mule, and they descended in six hours what had taken three days to climb. She saw a truck with a bright red cross painted on its side, and soon she was in a Bogotá hospital. Her life was saved.

The next day, Mark and Shen were also marched down to the valley where other FARC men were gathered. A Colombian army assault began, and the FARC men disappeared. The two Americans, some Colombian journalists and TV cameramen, and Red Cross workers ran the opposite direction. In a few days the birders were back home.

As I write this, Louise Augustine is in Peru, and all four of the birders have returned or plan to return to South America, but not to Colombia.

Todd Mark says, "You're vulnerable anywhere you go. Just because this happened in Colombia doesn't mean you shouldn't go anywhere."

***

There is another hazard for the birder that I have never seen discussed in the scientific journals. As I recall it, the scientific name of this condition is acute aposiopesis apotheostic paroxyic frazzletraum. The symptoms of this ailment can come on unnoticed, beginning with the pleasure of discovery. It progresses to delight, then to intoxicating joy. Suddenly, rapture sets in, and the

Tree huggers. In some parts of Oregon it could be dangerous to hug a tree. (Photo by Lew Adams)

Red inspects a big chunk maliciously cut from the trunk of this ancient giant by a logger's chain saw. (Photo by Lew Adams)

victim can't see enough fast enough. The victim's neck begins to stiffen, the arms grow tired, and the mind scrambles while trying to gather facts, visions, comparisons, movements, and sounds, all in double time.

Movements grow mechanical as the birder tries to be invisible. The birder sees an object of desire, but before he or she can embrace it, another has intruded. They move slowly, bending, kneeling, bending over backwards, and then race ahead, jabbering to passersby. They creep through thickets on hands and knees, unmindful of poison ivy, ticks, or snakes. They lie on their backs, trying to pierce the clouds. They are no longer conscious of their surroundings, for they have reached a dangerous heaven of shock and overdose.

This condition is most often seen in places such as High Island on the Texas coast during a bird migration fallout. My strongest advice to birders: Arrange ahead for someone to force you to drink water and eat food. Your survival may depend on it.

During my first encounter with this ailment in the Rio Grande Valley of Texas it hit me so hard that if Red hadn't rescued me, I wouldn't be here to tell it.

***

There is a term used by firefighters: risk-benefit analysis. Chuck Catt, division chief of the Austin Fire Department, clarified its meaning for me: "Risk little to save little. Risk a lot to save a lot."

Perhaps birders should learn to apply it too.

SNAKES

In the Texas Hill Country if you mention snakes, you're apt to get a conversation. There's no end of narrow escapes, and birders, of course, are among the escapees. Since I've mentioned snakes, here is a conversation.

There are fifteen species of venomous snakes in the United States, divided into four categories: rattlesnake, coral, moccasin, and copperhead. The land at the confluence of Roy Creek Canyon and the Pedernales River is home to all four. While I was living there, my greatest fear was always the rattler, for I'm burdened with the superstition that if one bit me, it would mean curtains. Thinking it over, I may be living on borrowed time.

Roy Creek always presented a logistical problem: Should we live in the easy-access ranch country on top of the bluff with the Bewick's Wren, the Common Poorwill, and the Black-throated and Lark Sparrows, or should we live in the lushness of the two-hundred-foot-deep canyon with the Louisiana Waterthrush, the Canyon Wren, and the Acadian Flycatcher? Through the years we did both, and either choice included snakes.

In our pioneering days more than fifty years ago, in order to live on top it was necessary to chop prickly pear cactus for a road and a good spot for the Road Roost, our twenty-two-foot travel trailer, which we were living in. Red drove thirty miles to work in town, and I stayed at camp alone without a car or a telephone or near neighbors.

I was busy chopping cactus and kept hearing an unfamiliar high-pitched sound close by. Finally, I worked near enough to realize that the sound came from a five-foot rattlesnake, mad as hell, rattling its tail to announce that this was its homestead.

I quietly sneaked into the Road Roost, got the pistol, and replied, "This is now my homestead, and, sorry, the two of us can't share it." Bang! Bang! Bang!

Mr. Rattlesnake's reply was not to roll over and give up but to make a speedy but dignified retreat to his rocky stronghold below the edge of the bluff. Since (in days of terrible ignorance) I once shot a flying crow dead with a .22 rifle, this was not the expected result.

"Of course you didn't kill it," husband Red commented matter-of-factly when he returned home. "I loaded the pistol with rat shot to get rid of the mice down in the cabin."

Yes, I did remind husband that our nearest neighbor was a long mile away, and I did ask him how this event might have turned out if the rattler instead had been a rustler who just happened to trespass and had a liking for lonely young ladies and any valuables he could carry off?

From down in the canyon it was seventy-eight steep steps from the cabin back up to the bluff top, and we almost always had too many items to load on the trolley. This time it was a bag of laundry that I would swing onto the step above me and then follow up. Just as I reached the top step, a rattler uncoiled and struck, stopped midstrike as it realized the laundry wasn't alive, then pulled back and struck at my leg. I did what any other fool would do and fell backward. So here I am—living on borrowed time.

I never passed the spot again without saying a thank-you and marveling at how a large snake could be so perfectly camouflaged in a few skimpy grass stalks.

This species' marvelous camouflage proved dangerous another time when we were filming and I sought a step-down on the edge of the bluff for a view of a red salvia. I placed my hand on a rock—oops! sorry! The nice little rock was entirely filled with a small rattlesnake. Lucky me! It was so early in spring that the snake was still sluggish.

In spite of my theories about rattlesnakes, like the folk writer–professor J. Frank Dobie, I would never kill the last one on earth.

I was working on a story about the cowboys who were patrolling our border with Mexico to prevent cattle from wading the Rio Grande. The dreaded foot-and-mouth disease had struck our neighbor country, and a stray cow or two could bring the dreaded plague into the United States.

Two of the range riders invited Red and me to join them that night to go fishing on the river. In the night blackness the path was a ghostly tunnel through tall, thick cane, and a cowboy named Lucky was walking too far ahead of Red's flashlight for common sense. Suddenly, there it was. An extralarge pipe—or was it a small telephone pole?—stretched across the path from cane to cane . . . a giant snake.

Lucky was carrying a discarded anvil, which was to be the anchor for the rowboat, and Red yelled a command, "Lucky, drop the anchor on it *now!*" Lucky was indeed lucky.

Somewhere we have a picture of me astride a pretty paint horse, holding

Marjorie holds a seven-foot rattlesnake, killed on the Rio Grande while she was covering a story about foot-and-mouth cattle disease in Mexico.

that rattler out by the tail with its head dragging the ground. It was such a granddaddy that it filled our ice chest to the top, and on the way home, on one pretense or another, we got the unsuspecting to open the chest as we quickly moved away to watch the show.

We don't like any wild thing to go to waste, and after I skinned the granddaddy, its beautifully patterned hide lapped around a twelve-inch board on each side. It was truly a Texas-size snake. He deserved to live, alas.

Since our Eden always had ranching neighbors, through the years cattle and goats sometimes got down into the canyon. Goats usually could get out, but we found a dead cow enough times to satisfy the need, thank you. It is hard to burn a dead cow, but burning it leaves the place more sanitary. Now, with better fences, cows seldom show up, but the last two came along while our grandchildren, Adam and Alethea, were visiting.

It was a noisy, rowdy, and strenuous event as the youngsters helped our neighbor drive the animals toward the only way out of the canyon. One of the cows tried to go into reverse, and Adam said, "Granny, I was running so fast and the ground was so rough, I couldn't stop. I didn't have any choice but to jump right over that big rattlesnake!"

In a half century we saw only two coral snakes. One was on the edge of the bluff in an area we seldom walked. It is hard to decide with this skinny snake species which is head or tail, as the mouth is small and the species is not aggressive. I watched the pretty little thing at my feet and let it go.

However, the other one was where we parked the cars, and we had small children with us. Our friend Sandy Richards took the snake and had a hatband made out of it. Coral snakes have a venom similar to that of the deadly cobra, so I guess that was the right thing to do with this one.

In a canyon as wild and hazardous as Roy Creek, if we went off alone, we made a practice of telling someone or leaving a note as to where we were heading. I explored many an inch of this rock-tumbled, precipitous gorge, much of the time solo, and was spared, you might say, on a number of occasions.

I was at Zilla's Spring, another of the natural swimming holes on beautiful Roy, and I wanted to see where the spring came out. It was (and still is) a very steep climb from the creek bank to the base of the bluff. I climbed and worked my way amidst tumbled rocks, little tributaries, ferns, and roots, and by the time I reached the bottom of the bluff, I was on all fours. The spring flowed out over a fern-covered shelf about four feet above me, and as I stood up at its foot, I looked directly into the eyes of a snake.

The large, dark snake lay comfortably relaxed amidst the delicate maidenhair fern. Our faces were not more than a foot apart, and I smelled moccasin something awful. I can't imagine what was going through the snake's brain as it suddenly had an extremely intimate look into the eyes of a human only a breath away, but I remember well what I was thinking.

The human tried not to blink her eyes, breathe, or give any other indication she was alive. She was sure the thump of her heart would upset the moccasin, and she intensely ordered the organ to shut up. She was absolutely sure there was no time to think about it, and when she moved, it was so fast that she began rolling and finally roll-bounced to the creek bank.

Sitting on a boulder, thoroughly bruised and poking and feeling around to see if I was still in one piece, I looked down into the crystal water of Zilla's Spring and began laughing, and I kept on laughing and laughing till I hurt. It was absolutely, completely, and utterly funny that I hadn't broken my neck.

Don't remember what else I was laughing about.

GETTING CLOSE TO NATURE

There were few of the many amenities available to today's campers, but in 1911, after a pleasant stay in a luxury hotel, my mother and father spent part of their honeymoon camping in a tent in the Texas Hill Country.

On at least one occasion Mother went with Dad on his two-week trip through sparsely settled West Texas. These were times when some "highways" went through fences from one property to another by way of cattle guards. These innovations, which augmented gates, were originally intended for wagons and buggies and constructed with a middle support so high that only Model T Fords could clear them. Dad always carried extra oil, water, and gas in containers on the running board of his Model T.

Part of the time on such a trip my folks would camp out, and once they set up their tent in a pretty spot on a bank of the Rio Grande. It was not uncommon in those days for shots to be fired back and forth across the international border, and occasionally somebody got killed. My nature-loving parents spent an uncomfortable night with sporadic shots ringing out from Mexico.

I began to experience the outdoors early in life because my poet-artist-mother found great joy in painting landscapes in the countryside. I picked armloads of wildflowers, learned you're always closer to prickly pear than you thought, and gained firsthand knowledge of ticks, chiggers, and poison ivy. Also, it is wise to keep a sharp lookout for bulls.

Oh, that feeling of luxury when I lay down in a bed of clover!

When I was about twelve, I had my own garden patch, which I planted with wildflowers from neighboring vacant lots. I also used box-crate lumber to build myself a lookout in the fork of a large hackberry tree. I would come home from school, get a biscuit left over from breakfast, put an onion from our garden inside it, then climb into my aerie to eat and to contemplate the world, feeling, I am certain, the same power of domain that a Golden Eagle might feel when perched on its favorite high crag.

A special summer camp for girls was located in the countryside, but I don't recall any specific attention there to the wonders of nature other than a hike through the woods and cooking weenies by the campfire. Foremost was the never-ending chatter about boys.

Husband Red and I began our camping adventures with old quilts spread on the ground. These were roofed with a mosquito bar draped across a line stretched taut from a stub driven into the ground at the head and another stub at the foot.

We cooked on campfires, and I became an expert fire builder. Old bacon grease burns well. I have been known to successfully bake biscuits in a heavy cast-iron Dutch oven buried in the coals.

During a Christmas holiday in about 1950 we loaded up our serviceable Dodge sedan with bedding, tarps, and food box; tucked away pistol and our kids, five and eight; and headed out to Big Bend. With a blue Texas norther blasting our backs, we were pleased to spend our first night in a deserted WWII barrack.

Though Big Bend had been declared a national park in 1944, we found no sign of this anywhere in this great western desert, and the only people we saw were John Skinner, a cowboy; his wife, Cornelia; and Shaw, their toddler son. They were living in a big tent near the river. John's job was to patrol the border for a designated number of miles daily to prevent Mexican cattle from crossing the Rio Grande and possibly bringing with them the dreaded foot-and-mouth disease. The three were so remote from everything, anywhere, that they were especially pleased with our company. It was about 150 miles to the grocery store.

It was convenient for us to camp nearby at the hot springs, which were enclosed with a woven cane windbreak. We had all the hot baths we wanted and, of course, plenty of hot dishwater.

Our primitive camping experiences ended when we installed a canopy manufactured to fit the rear of our new truck. Inside this rainproof shelter, Red built a double bed. We couldn't stand up in this outfit, and we cooked outdoors on a small gasoline stove and used a gasoline lantern for light.

Returning from the American Birding Association's first convention in North Dakota, I remember eating breakfast with gloves on in June at Fish-eating Creek. It was only later that we noticed the nearby snowbank.

Our next step was a crew-cab truck, which could hold six people. A large camper, part of which extended over the truck roof, slid compactly into the truck bed. This abode had a rear door, roll-out windows, a propane stove and refrigerator, a sink with one faucet, a dining table and seats that were convertible to a bed, gas and electric lights (when a hookup was available), a tiny

Camping with kids on the way to Big Bend, Christmas holiday 1950.

The Adams family, enjoying a riverside lunch along the Rio Grande with border patroller John Skinner and his wife, Cornelia, and son, Shaw.

Luxury camping on the Texas coast with A. D. and Jean Stenger. Inset: *Marjorie and Red with some of the day's catch. (Courtesy of A. D. and Jean Stenger)*

closet, and a double bed built over the roof of the truck, which Red declared "needed a running jump to get into."

We spent a whole summer in the truck while filming in Yellowstone National Park, but we were hampered by the fact that I broke a wrist. This happened when we were preparing to film out on the ancient Bannock Indian

The Adams pair, rafting the Rogue River, Oregon. (Photo by Lew Adams)

Trail. Loaded with film equipment, I stepped in a "whistle pig" (ground squirrel) hole. The effect was similar to being thrown from a horse.

With me patched up by the visiting park doctor, we stayed on and Red got footage of a grizzly. It wasn't till we got home that we realized the rowdy noise that awakened us one night had resulted in the imprint of a grizzly's paw in the camper's metal wall beside the back door. We were thankful, belatedly, we had roused enough and talked enough to send the bear on its way. Undoubtedly, the grizzly had smelled the fresh cantaloupes we bought that day.

Along with thousands of other Americans, we promoted ourselves to more luxury in the 1960s, with a twenty-two-foot travel trailer. Our parents had died, our children had married, so this time we lived in it. Its permanent address was on the Pedernales River, but home also could be anyplace—a national park, a lakeside or seashore, or, if we were working on a film, a trailer park in the middle of Los Angeles.

Here's a paragraph from one of my columns dated December 1972: "For more than seven years I have written Bird World at a small desk-dining-vanity table, and my desk seat also serves as a sofa–guest bed. . . . The Road Roost has been not only a tool for adventure but an instant, anywhere living room to which birders and nature lovers of all ages and aims from various states have come to visit, rest, or eat. It also has been a natural blind from which we have watched all kinds of birds in all types of habitats in all kinds of weather."

Red and Marjorie filming in Yellowstone National Park despite Marjorie's broken wrist.

For instance, we parked on a bluff overlooking the Pacific and, in a rainstorm, were able to watch a sea otter calmly floating as it used a rock to crack a shell resting on its chest.

Finally, we gave in to a twenty-two-foot motor home. Though in Red's words "it needed a lot more room than a stick horse," it still could be parked in a regular parking place. We never pulled a car behind it, so if we were parked and connected to water, lights, and plumbing, we'd have to disconnect every-

*Field trip to the Davis Mountains with Ed Kutac and a group from the
Texas Ornithological Society. (Courtesy of Ed Kutac)*

thing to run an errand. Despite this inconvenience, there were tremendous
advantages. One of us could rest while the other drove, and the last time we
headed for Mexico, I baked a smoked turkey as we traveled, and it was ready
to eat soon after we crossed the border.

In Mexico and Belize the motor home was regularly greeted as a "happening" in towns, and in the more remote areas it always gathered a crowd of
people who obviously had never seen anything like it. Some were fearful of
it, but others marveled. I will never forget the teenaged señorita who looked
around and gasped, "Es una casita!" Then she began going through the motions of cooking on the stove, putting the dinner on the table, and sitting
down to play-eat. If we had charged admission for the tours we conducted of
the Road Roost, we could have paid for our trip and had money left over.

Living in a space eight by twenty feet is instructive. The primary lesson is
how little we humans need in order to live a comfortable, safe, and produc-
tive life. It teaches multiple uses of space and, to the utmost, how to store
and arrange possessions, including books, files, typewriters, cinematography
equipment, and fishing rods. I had one party dress, which folks could just
get accustomed to. And, of course, our travels required at least some of both
winter and summer apparel because of changes in altitude.

Living successfully in a small space also gives one perspective, as well as an attitude toward extra-large luxurious mansions that borders, I freely admit, on the judgmental. Taste, beauty, and convenience don't depend on money. I agree with the American Indian chief who said, "Take only what you need." Many of the miles we covered were marred with a thoughtless, man-made ugliness, which is spreading, in those knowing words of another observer, like a "planetary disease."

Red and I took turns driving, but when we were in cities or following directions and maps, Red drove and I navigated as we wandered from Canada to Belize and from Miami to Los Angeles. I boast that we were a good team (and certainly extremely lucky) as we experienced hurricanes, downbursts, earthquakes, floods, ice-overs, close-shaving trucks, freeway spaghetti bowls, dead ends, smog alerts, sand traps, and much, much more at altitudes from below sea level to nearly twelve thousand feet.

The thrill came again and again from the earth itself. We experienced thousands of habitats (and the wildlife they supported), from mountain heights to jungles and seashores, and, in all, they included five languages, two of them Indian.

There is no show more marvelous than planet Earth. How I wish we could start out again in the morning!

* * *

But no matter how many the miles nor how great the allure, we always came back to a patch of land in the Texas Hill Country.

It was not the most sensible place to settle down, a terrain made up of hundreds of alternate elevations, but enough of it was level that we could park a twenty-two-foot trailer. In 1968 the telephone line had finally inched its thirty-mile way from Austin, water was being piped 130 feet up the bluff from the spring far below, and Lyndon Johnson had gotten electricity for us rural Texans. Water, lights, and telephone—what else could one want in a place with this much natural beauty?

Friends had begun to call it Eden—a crystal-clear spring-fed creek with dozens of waterfalls and pools of many sizes, all bedecked with maidenhair and other ferns and guarded with giant cypresses bearded with long strands of gray moss—this, and the river with good fishing just a stone's throw away.

Yes, a true Eden, but it's all at the bottom of a two-hundred-foot canyon. No matter. Marjorie and Red are young, the steep trail is climbable even with a load, and it *is* an all-beauty dream. Let's call this place home, at least for now.

The Texas Hill Country, is a land of plateau uplands, ruggedly dissected limestone hills, thin and stony soils, and juniper and oak forests, all lying west of the Balcones Fault. It has hundreds of springs feeding crystal-clear, cypress-lined creeks. Roy is the name of the creek that graces Eden.

Roy Creek has dug itself a canyon and in the process sloughed off cliff parts and boulders, some larger than a house, and strewed them below in confusing grandeur. This particular parcel was just under 25 acres, a land pittance in Texas, but after exploring its ups and downs, an engineer once told us that if this uncommonly rugged property could be rolled out flat, he estimated it would comprise at least a section, that is, 640 acres.

This part of Texas is a major weather-maker. The Balcones Fault is the first topographic break in elevation coming west from the Gulf of Mexico and thus is an area influencing unstable, water-laden air masses. The Balcones Escarpment is the locality of some of the largest flood-producing storms in the United States and, in fact, the world. The greatest single rainfall ever recorded in the United States occurred in 1921 when thirty-eight inches fell in twenty-four hours near Thrall, Texas, a record that stood for fifty-eight years until Hurricane Allison came along in 1964 and dumped sixty-four inches on the Houston vicinity.

Besides the extreme of floods, this unique land must also endure the extreme of harsh periodic droughts at the same time it is fabled as the land of ten thousand springs.

Thankfully, we were not at Eden when a total of twenty-three inches of rainfall from two watersheds caused an eighty-foot rise on the Pedernales River, an event marked to this day by a steel rod that Red drove into a rock halfway up the high canyon bluff. This flood snapped off the trunks of cypress trees five feet in diameter as if they were matchsticks. It would also have swept away the primitive cabin down in the canyon if the cabin hadn't caught on a protruding boulder.

Anywhere along the cliff's edge at Eden two sharply contrasting worlds meet, the verdant, closed-in canyon and the semidesert plateau. Standing at the bluff's drop-off, we could hear from below the sound of the waterfall, the songs of the Red-eyed and White-eyed Vireos, and even the hiccup of an unusually far-west Acadian Flycatcher. Their world is in an envelope of trees, a world where the morning sun takes its own sweet time appearing above the bluff.

On the banks of Roy Creek, white-trunked sycamores contrast with lacy black walnut trees, and bald cypresses up to six feet in diameter tower toward the cliff top. For generations the stumps of even grander trees remained as reminders of the sojourn of the Mormons, when they harvested and lumbered the ancient giants and then floated them downstream on spring floods.

The centerpiece of Eden is a twenty-foot waterfall splashing into a clear

Close-up of Venus's Bathtub. (Photo © Lew Adams)

and cold nature-made swimming pool encased in enormous boulders graced with maidenhair ferns. Highlights and decorative touches are supplied by wild plums and redbuds and the nest of the Eastern Phoebes, which the birds had placed in the same rock indentation under the waterfall for at least ten years. We called this pool Venus's Bathtub, and we were certain the goddess luxuriated in it at least in her dreams. Or was that the vision of our daughter, Zilla, floating with her beautiful red hair?

In contrast, up on the sparse plateau, the sky is major twenty-four hours a day, from the dawn's glimmer to the lingering sunsets, sometimes with massive thunderheads, and then the night grows into a spacious moonscape so clear that objects stand apart from their shadows. And, oh, that delicious summer-night breeze, which below is blocked by the bluffs!

Red and I did most of our traveling and filming in our travel trailer, but now we were also living in it. Not even a parachute could successfully place this abode into the canyon in one piece, so the Road Roost was parked on top of the bluff among shaggy, very old Ashe juniper trees and mistreated live oaks out on what we called Rattlesnake Point, or just the Point. This spot, which overlooks both the creek and the river, includes a big chunk of the bluff, a limestone island that has split off and stood apart and steady for at least a century or so.

The chasm formed by this separation can be crossed with a long jump, but it was more safely spanned by two juniper logs topped with boards, a small, neat bridge built by Red as a gift to me. As we stood on this Point, looking down at the blue-green Pedernales, we would tell visitors that we swam in the river after President Lyndon Johnson walked on it upstream.

The Point was also where we sat on two old-fashioned lard cans to watch the sunset and consider our lives and blessings or even the lack thereof. The blessings always won.

As we settled in, we took up patterns and habits the same as those of our old friends who had been here long before us. As far back as I can remember, there has been a Canyon Wren at the Point. This is where we could look down on the riverbank and see sunflowers, milkweed, that dreadful, stickery-mean bull nettle, and even castor beans. Sometimes there's a Painted Bunting there or even a Blue Grosbeak.

On the new power-line pole installed by the Pedernales Electric Cooperative (thanks to Lyndon), the Summer Tanager sang part of the time and the Ladder-backed Woodpecker drummed regularly, just as they had last summer. We could still hear the Belted Kingfisher as it flew up-creek each morning, rattling. Every daylight a turkey gobbled across the creek. At 7:20 each evening the Common Poorwill started his lament from the Indian-Grave Hill. The Cliff Swallows were under the same cliff they may have used for a hun-

Friends enjoy John Henry Faulk's storytelling at Roy Creek.

dred years. The Black-chinned Hummingbird still favored the same cedar snag for perching, restlessly on the alert for intruders. Even that very special fellow, the Golden-cheeked Warbler, buzzed his faint *wee-ah-weezy* song from the same tree branch where Red filmed him.

Some birds seem equally at home in either world, and we had scarcely parked the Road Roost before the Bewick's Wren came peeping in the windows just as it peeked in the windows downstairs. Upstairs and downstairs the clicking female Brown-headed Cowbird skulked, invading nests in cedars and scrub oaks as judiciously as the nests in white oaks, buckeyes, and cypresses.

The most cardinals I have ever seen in one spot were at the entrance to Santa Ana National Wildlife Refuge, so we designated a section of our neighbor's ranch road, where tree branches met overhead, as Cardinal Arcade, because the numbers of cardinals there almost equaled those at Santa Ana's. (Later, one of the various succeeding ranch owners cleared everything but a few oak trees—farewell to Cardinal Arcade.)

Lark Sparrows nested nearby, and directly over us the tiny lichen-covered nest of a Blue-gray Gnatcatcher stuck to an oak limb. A young Field Sparrow in first nuptial plumage was trying near and far to sing up a mate. At dusk Common Nighthawks flew a patrol up and down the white caliche ranch road. If you were there, they might come so low that you were sure they were going to swat you. We saw Northern Mockingbirds only around the ranch barn.

Mama and Greedy, the white-tailed deer, came systematically for their evening corn. The jackrabbit, which Red christened "Briptus" (a name he couldn't explain), ate what was left of the grass, the bullfrog boomed annually from the riverbank, and so it went. Like us, animals and birds have their daily routines.

But things aren't really the same. The Eastern Bluebirds disappeared when their nest tree fell. With the disappearance of two giant prickly pear cactus plants, the trusting Black-throated Sparrows moved farther away. Too late we knew our guilt and wished with all our hearts that we could put their cactus nesting place back.

The groups of Brown-headed Cowbirds showing up on the new telephone wires undoubtedly contributed to the absence of the Black-capped Vireo, which I had stalked and stalked and stalked some more in the thickets here before I finally got a solid sighting. As good old Voltaire repeated: "Nothing is permanent but change." And, as always, some of the change is good, and some bad.

A friend had now declared that Red and Marjorie Adams lived at the Innermost Outpost. This could be true. Our livelihood was thirty miles away, partly by a dirt road, which in wet weather liked to grab cars in deep ruts or slither them askew. Our mailbox was two miles away, and the nearest neighbor almost a mile. Isolated from conveniences and easily available hired labor, we might not make it, but Eden had beckoned, and isn't Eden what all of us are seeking?

We were prepared: our part of the new telephone line that we would share with three other parties could be turned off.

* * *

After the church service on April 29, 2002, a visitor came up to me and pointed to my name tag. "So you are Marjorie Adams." he said. "I'm glad to have a chance to visit with you. Do you remember my wife, Martha Watkins?"

"Yes, of course I do. I especially remember Martha for her love of the outdoors and nature." Then I asked the question regularly asked by elders my age. "Is she still with us?"

"No, she's been gone several years now. Seeing your name reminded me she visited with you out in the country years ago when you and your husband first moved there. She told me how beautiful and wild the canyon was, and she was much impressed with your courage.

"In fact, I'll go ahead and tell you exactly what she told me. She said you and your husband were considered the hippies of those days."

We never know how we'll be remembered, but "hippies" suits Marjorie and Red Adams just fine. And remembering the long-ago makes me wish, Oh, to be a hippie again!

THE THRONE

The cabin that was long called Red's Roost sits on a shelf that extends out from the canyon wall about halfway down to Roy Creek. This shelf is about three hundred feet long and about sixty feet at its widest. When you exit the cabin's north door, a turn to the right would start you, in about fifteen strides, on a climb back up the eighty-five-foot bluff. A turn to the left outside the door would place you in a stride or two down the forty-foot bluff into Roy Creek. Going straight out the door puts you on a short trail that leads through a small grassy area with scattered elm trees, and then the path skirts a jumble of huge boulders tipped on their sides. They are at least fifteen feet tall, and only the Devil knows how deep they go into the ground. In ages past they could have been the ceiling of a grotto through which Roy Creek flowed.

At the end of this row of gigantic stones is a bell hanging from a chain. It is the custom to ring it to ascertain if this place is occupied. If there is no answer, you then turn to the right to walk around and behind the giant stone walls, and there you'll discover a small hideaway enclosed on the other side by the steep and crumbling bluff. Sitting there in isolation is the Throne.

Deep in this canyon, Red's Roost has hot and cold running water and a sink, an electric cookstove, a large refrigerator, air-conditioning, a radio, and a telephone, all indicators of civilization. But where is the bath? The creek works just fine most of the time, and for a cooling shower before bedtime you can catch a direct flow of frigid springwater at the outdoor kitchen sink.

That's it. There is no sensible way to install a septic system on this shelf of rocks, so the solution has been the Throne.

There is a half-true joke in the Hill Country of Texas that the hardy cedar choppers and charcoal burners who have inhabited this area for generations have only one piece of furniture, a highly portable board with a hole in it. Therefore the Throne has a long ancestral history behind it. Ours was a neat shallow box that sat firmly on short sturdy legs cut from the enduring trunk of a juniper tree. There was a comfortable hole cut in its solid top.

The Throne was an open-air privy with a view—no roof, no enclosing walls except those supplied by the rocks and cliffs. The natural question could be, Why, oh, why haven't you folks improved this situation in all these years?

Ah, if only you knew how much time and thought had been applied to the Throne. There have been long explorations of such things as sump pumps, which can pump all the way to the top of the bluff, but the terrain there is even less favorable for disposal than the terrain here. Then there are such gadgets as gas- or electric-powered toilets used by the military in remote or formidable locations—expensive and certainly not foolproof nor maintenance-free.

How do you get a plumber to come thirty miles out of town to work in a deep canyon? And if you did get the plumber, what would he or she do that would be any different from what's here?

Besides, Red and I were the only inhabitants of Red's Roost, and we never lived there in cold weather. A sprinkle of lime with perhaps a cover of leaves and a paper sack on the side to hold used paper, which would be burned, all worked well.

We once planned to install a cane screen or something similar for the sake of privacy, but digging a posthole to support such screening can be not only arduous but also disappointing, for perhaps less than a foot down the digger encounters solid rock. And so on. Besides, much of the time Red and I were living in the motor home on top or heading off to California, Canada, Mexico, or Austin.

Finally, the Throne, as humble as it was or perhaps because of its humbleness, began to achieve a sort of status. Marshall Johnston, our botany expert, went so far as to say it had acquired "a certain charm." For city folk it came near to being an Experience that not only was a surprise or a shock, depending on the visitor's personality, but also something to be remembered, joked about, survived, or even revered.

The Throne made up for its lacks with its intimate views of its surroundings. This panorama seemed scarcely to have changed a limb, a leaf, or a rock in the half century I observed it. This is a land that grows sedately and carefully to plan for the sure-to-come drought.

Rightfully designed to grow on top of the bluff, a twisted-leaf yucca persisted there, though it never bloomed in the shade. A Mexican buckeye lasted long enough to become a small tree, and in a particular fork of a particular branch shaded by a small branch above it a White-eyed Vireo had a nest every year. I believe you could count on it, even when the cowbird found it too.

As in areas with true mountains, the bluff has its own flowing rock stream, which to the casual eye seems stagnant, yet rock chunks still roll down in all sorts and sizes to lie in ordered disorder from season to season. There are many kinds of lichens growing at Eden. The rocks at the Throne are shaggy with them, and they always bloomed to our notice after a rain.

At least a dozen different kinds of ferns survive at Eden. Some of these cling to the shady side of the rock wall. The resurrection fern curls up and dies when it gets too dry, then unfolds to life when it rains. Spring will bring the scarlet of the cedar sage, a name I think is far too lowly for the plant's delicate stalks of blazing blooms. About four feet from the Throne, a rock about a foot in size had a different species of moss on it that turned the brightest velvety green with moisture. It was a small verdant jewel that stood apart in a rugged, rough place.

It was spring as I sat on the Throne, contemplating these familiar friends. Suddenly I had company. A small bird caught a caterpillar from the buckeye on my right, then lit on the verdant rock less than a yard from my feet to worry the fat feast before dining. It was my habit to take my binoculars with me on these visits, but I certainly didn't need them with this bird. I knew at once it was a Lifer.

There are many different kinds of ants in the Hill Country. One of them is very tiny, black, and related to fire ants, a species noted for its sting. As I gazed in a trance at the bird, at little more than an arm's length from me, it was at this moment that the vicious little ants awoke to a territorial dispute.

As the first bite stung, I sat resolutely motionless and merely blinked as I studied this bird, a gift handed to me literally right under my nose.

It obviously was a very weary bird in migration, and it chose now to rest and preen in this quiet place. It had a vireo bill and undoubtedly was a vireo, but of course not the White-eyed nor the Yellow-throated, and it was too calm and late for a Red-eyed.

At this time a second ant got angry with me. Its bite was suitable for an ant at least five times its size. I breathed harder as inconspicuously as I could and concentrated on the bird. It was so pale, so nondescript—I needed the field guide.

Soon it was apparent that I was blocking an ant trail. Traffic was building up. So were the ants' tempers. The stings multiplied, and each was like a hypodermic of blistering fire. I ordered my best self to come forward. "When did you ever have a Lifer almost sitting in your lap?" I thought. *"Concentrate!"*

I clenched my teeth so hard they could have cracked. "Don't even think of scratching," I told myself. "And you'd better stop holding your breath— you might fall over."

I was imprinting this bird on my brain in competition with powerful other matters, such as that I might scream with the next bite. "Yes, a faint little eye line but no wing bars," I said to myself. "In fact, this is a pale copy of a bird, just gray above and whitish on the breast. Is this not-so-pretty bird worth what I'm going through? *It's a Lifer. Concentrate.* Where's your guts, you dumb-dumb?

"OK," I resolved. "I will sit like an Indian in purdah, doing penance. I'm not moving. This bird is going to move first, even if I die here—all alone perched on the Throne. Sounds like the first line of a little verse, which I will not write. Remember, I'm not giving in. No. Not unless I go crazy-wild."

I won! The bird moved first! Then it flew away.

Without taking even a millisecond for ant revenge, I fled to the cabin to apply alcohol, calamine lotion, and ice and finally to swallow a pain pill. Those stings, and there were many of them, itched for nearly a month.

But I'm certain beyond doubt that the memory of my first pale, petite, shy, greenish gray Warbling Vireo will last a lifetime.

And the Throne endures.

BEDI THE BIRD MAN

It was the early 1950s, and this was to be my first interview with Roy Bedichek. "The best place to have a heart-to-heart talk with me is in the back of a garage," he said on the phone. "It's the dirtiest place you ever saw," he warned. "I don't clean it up except every now and then for the visit of a very respectable person."

"In that case, of course, I'm expecting it to be dirty," I said.

"When you get here, the latch string will be out for you," he instructed. "Just pull the string and come in."

We had talked on the phone several times, the first at the urging of a friend: "Why don't you call Mr. Bedichek? He knows all about the birds, and I know he'll be glad to help you."

The children and I were spending the summer at the cabin in Roy Creek Canyon (with Red there on Wednesdays and weekends), and I was seeing birds in quantity daily. My frustration was huge. The book I was using had cost a dollar secondhand, and it was worth every bit of a dime. It had itsy-bitsy black-and-white drawings of birds, often shown on their backs, feet in the air, and these "illustrations" were accompanied by tedious scientific descriptions.

Bedichek's acclaimed book *Adventures with a Texas Naturalist* was published in 1947, and the author had continued to gain stature and was even compared to Thoreau as an observer of birds and nature. So it was with considerable timidity that I finally called the great man. He attentively listened to my description of "this tiny bird that flutters so much I can't keep up with it" and said, "Why, Mrs. Adams, you have a kinglet."

Bedichek was more fortunate than I in his bird study. About 1914, while he was serving as secretary to the president of the University of Texas, a man named William Taylor began teaching agriculture there. Taylor was intensely interested in birds, and when Bedichek began sharing walks with him, he also shared Taylor's bird knowledge. At that time this was the best way to learn birds, for there were few books available.

Now I was standing at Bedichek's garage door, an inexperienced, "un-birded" reporter—and awestruck. When I pulled the latch string, I would be face-to-face with the noted bird scholar. Finally I pulled, the door opened, and there stood Roy Bedichek with a dust cloth in his hand.

The place was clean, and cozy too. He hadn't mentioned that he had re-modeled this particular garage into a retreat lined with books, files, and me-mentos. A large potbellied stove reigned in the middle of it all.

"Have you ever seen how the latch string works?" he asked.

He had bored a hole in a perfectly good door through which a sturdy knotted string hung to the outside. "See?" he said, as he demonstrated for me. "By pulling the string, you raise the inside door latch and the door will open. When you want to lock up, you pull the latch string back inside and the door can't be opened." He had made me at ease.

"I saw your martin house out here," I told him. "My husband and I put up a house but never got martins to come as far west as Roy Creek."

"Well, I've had 'em for years." As we watched the industrious birds fly in and out of the nest compartments, he remarked, "Martins don't sing; they have conversations. They make one of the most cheerful of all sounds in nature.

"I made this sixteen-apartment house down at the lumber mill eight or nine years ago. I got one of the holes a trifle too large, and a screech-owl moved in. There wasn't any conflict between it and the martins, because the owl built at night and the martins built in the day. However, the martins began to bor-row the owl's building materials. Papa Martin would steal a piece of grass or a stick and hand it around to Mama at their own doorway. She would measure it, test it, and weigh it, deciding if it was good enough. If it wasn't, she'd drop it on the ground. I don't know if the owl ever retrieved it when its building turn came at night.

"I guess you know several martins teaming up can air a hawk clear out of the country. That's why farmers keep martin boxes around their hen yards.

"I used to think my martins came back each year on March 2, Texas Inde-pendence Day. However, after six straight years this charming tradition was nullified, as they returned on various dates, even as early as February 15."

He stroked the leaves of a nearby plant as if it were a well-loved friend. "I don't have anything but native plants around my house. This is a cockspur hawthorn," he said as he inspected it closely. "It had an illness last year, and I thought it was dead. I was just about to dig it up when the thing put on a few leaves. After more rest through the winter, it came out again in beautiful foli-age clear to the end of the 'dead' branches. A folk name for it is 'kimikinich,' and it's also called deer laurel."

There were birdhouses in trees and on poles. "There's a shortage of nesting places," Bedichek explained. "I'm trying to help out."

Bird study was always a large part of Roy Bedichek's long life, but he had also been a cotton picker, a schoolteacher, an assistant to a fake cancer healer, a homesteader in a dugout in Oklahoma, a berry picker in New Jersey, and a newspaper reporter, editor, and publisher. In 1908 he rode a bicycle from Eddy, Texas, to Deming, New Mexico, to become the secretary of Deming's chamber of commerce. He was also a laborer on riverboats, a slaughterer in a packing plant, a waiter, and a coal miner.

His real career was with the Interscholastic League of the University of Texas. For thirty years, much of that time as its director, he helped the league grow until it was the largest organization of its kind in the nation, with a membership of 2,300 schools and 400,000 students.

How could a man with this demanding career find time to explore the wild world in such detail? Early in his work with the league, Bedichek began traveling eight thousand to ten thousand miles a year. That's when Bedi began to camp. He had two sets of efficient equipment, one for camping and one for the city. Stopping beyond the outskirts of the cities, he would spend the night in the open, having a camp supper and breakfast.

"The pioneers had to camp in the valleys to get good water," he told me, "but I carry my own water and a little gasoline stove, and I find the highest confounded flat-topped hill around and cheat the mosquitoes."

In the early mornings he would heat shaving water on a portable stove, slick up in his good clothes, and arrive in town ready for work. The system gave him precious time in his beloved outdoors, and besides, he could pocket the money a hotel would have cost. On all his camping trips he took a satchel full of books.

Through the years he wandered some 300,000 miles around the Lone Star State, and his daily lists of birds, plants, and wildlife observations became a nature journal that formed the basis for his books *Adventures with a Texas Naturalist* and *Karankaway Country.*

Those books and many of the letters of his voluminous correspondence were written on a 1915 Oliver typewriter, which was still on his desk and in use. Those letters were often treasures to those who received them.

There was a portrait of Thoreau on the wall and beneath it the words "In each dewdrop of the morning lies the promise of the day."

Bedichek's writings have been likened not only to those of Thoreau but to other master nature chroniclers as well. Certainly, Bedichek brought to his writing years of study of the classics, both prose and poetry, much of which he could call forth from memory. One of his closest friends, J. Frank Dobie, himself an honored Texas writer, said, "Bedichek has the most richly stored mind I have ever met; it is as active as it is full." Stanley Walker, writing in the *Saturday Evening Post,* noted, "Some connoisseurs of the state of culture . . . rated Bedichek the most civilized man in Texas; he has been called the most natural man, one of the half-dozen best educated, and the most persuasive."

Walter Prescott Webb, professor, author, and historian, was the third member of this intimate triumvirate. He said, "Bedi's real loves are literature, nature, and conversation. To quote my good colored friend, Abe Cartwright, 'Mr. Bedichek is the best single-handed talker I know.' "

Years later I was told by someone who had been around all three men that Bedichek was the leader of the threesome: "The other two respected him more than they did each other."

In those early years the chill natural waters of Barton Springs swimming pool was the only air conditioner for many citizens. The three men were among those who liked to lounge on the banks, getting back in the water from time to time for a fresh cooldown. There was a small tree near the water on the far bank, and Mr. Bedi favored its shade. My redhead's skin also favored it, so I sometimes shared a visit with Bedichek and Dobie. I don't remember Webb there.

In 1995 a bronze, almost life-size sculpture of the three men as they would have lounged together on the bank of Barton Springs pool was dedicated and placed near the pool entrance. It is now known as *Philosophers' Rock.*

Bedichek's ability to quote from the great writers and poets was well known, but Bedi also had a reputation for wit and jokes. When I asked him which books he would take with him if he were stranded on a desert island, he was reminded first of the British actress who was asked which man she would most want to have with her on a desert island. Without hesitation she answered: "An obstetrician."

In regard to his own book choices for a desert island, Bedichek, instead of giving a list, said his first act would be to try to record from memory all he could of the great classics. This would be a goodly amount of Browning, Shelley, Tennyson, Wordsworth, Whitman, and other poets, besides many of the prose classics. After he had assured himself of this library, he would then take time to describe the daily happenings and experiences on the island.

Years before the game and sport of birding would be widely popular, Bedi seemed to predict the future when he wrote, "Joy shall be in the bird-lover's heart over one new bird more than over ninety and nine already listed. If the newcomer is found to be nesting in territory well outside his usual breeding range, the event stirs the amateur still more deeply."

He also told me, "Identification should be first. Remember, one of the chores God gave Adam was to name the animals, and, of course, that included birds. You can't really know a plant or a bird until you know its name. Even a distant star has a friendlier twinkle when you know its name. But rivalry in identification can be carried to an extreme."

It would be interesting indeed to have Bedichek's opinion of the growing competition in birding. Since a major purpose of the Interscholastic League was to foster competition to be the best, I think Bedi would laud competition and sport, but without doubt his first and foremost thought always would be protection of the birds. His opinion of the increasing commercialization of the birding game might be less approving.

He regretted the lack of affection in everyday life: "We have become dominated by a cult of unemotionalism. We speak of 'cold' scientific fact as if temperature had something to do with verity. . . . But surely only the phlegmatic person, professional or amateur, can see the vermilion flycatcher for the first time without a gasp of surprise and pleasure."

It was typical of his thought processes that this pleasure led to a search for why the bird had appeared in Texas, "flown here from the Tropics on its own power," and was apparently extending its range northward. His research showed it was the federal government's subsidy for construction of well-placed permanent earthen water tanks that gave the Vermilion its desired habitat of semidesert with accompanying water. By the 1940s thousands of such tanks had been built, creating brand new habitats for hundreds of forms of life besides birds. Bedichek cited that in four counties alone in the semiarid Edwards Plateau nearly eight hundred acres of water surface had become available, with such immense implications for increased wildlife that they could not be estimated. As a side note he wrote, "The only good thing I ever heard about a South American dictatorship is that the South American form of the vermilion flycatcher is protected in Argentina by presidential decree."

Bedichek made friends with hundreds of landowners, sharing their stories and love of nature along with his own expert knowledge—thus engaging them in the husbandry of their land and wildlife for future generations. Since in Texas most of the land is still in private hands, organizations and governments are following exactly the same policy today as they coach landowners in husbandry for future generations.

In the role of reporter, I felt freer to ask Roy Bedichek questions than I otherwise would have. Certainly, conversation was one of his great loves, and I was guilty of stretching our visits out. There was always something new to see or learn. For example, a real garage attached to his study was full all around the walls with camping equipment, everything from a flower press to a new pickup truck with a detachable canopy, which was dropped onto the truck by means of pulleys.

"Did I ever tell you how to read a classic?" he asked one day. "A classic must be absorbed, and this is most readily accomplished beginning about four o'clock in the morning. Get a pot of coffee ready, place a huge dictionary on the table, prop the classic against it, and have your pencil and notebook at hand. Drink the coffee slowly as the classic is absorbed—about an hour and a half to a cup and a half. Of course, the coffee may have to be reheated in that interval."

In the afternoons, Bedichek napped, visited with friends, or worked in his year-round garden. "I make my own soil for the garden," he told me. "The neighbors give me all their oak leaves. I don't use anything for fertilizer but sheep manure. Never use pesticides—they kill all the beneficial organisms, not to mention the earthworms."

The master's garden was truly a sight to behold, not overly large but every inch a treasure. We sat on its raised wall as the late sun intensified its lushness and Bedichek gave me the privilege of hearing him bring forth from his memory jewels from the poets. Then he faltered. The next line wouldn't come.

For a brief time he was an old man. He rose and swept his hand around the garden and laughed. "Well, I can remember Whitman," he said. "As for you, corpse, you'll make good manure."

He cut off three of the leaves of a beet plant. The well-fed giant rose three feet in height, and its deep green leaves, scored with red veins and stems, made a richly colored armload bouquet.

As he handed them over to me, he said, "Remember that old saying, 'The latch string is always out for you.'"

Bedichek had been dead about six years when I talked one day to his widow, Lillian Bedichek. "I always had better eyes than Roy," she said, "and when we'd go out to look at birds, I'd often see the bird first without binoculars. Then he'd get me to go around the other side of the bush or tree and scare the bird out so he could see it. Oh, he had a way of seeing birds!" She laughed.

I also had a talk with Bertha Dobie, Frank's widow. "What I remember most about Roy, of course," she told me, "was his love of nature and of birds most of all." There was a long silence before she continued. "He and Mrs. Bedi were a little older than we were, but we all had the same maturity, so we were really of the same times.

"I don't know whether this will be useful to you at all, but he stood in front of our fireplace one day—I can just see him. He said, 'We were born in the best of times.' Then he continued, 'It was the best of times. We knew what is certain. Now people don't know what is certain. I know I'm not as happy as I once was, but I don't believe I see people as happy now as we used to be.'"

Was that because we are living in a world increasingly far removed from nature?

SCOTER SCOPE-OUT

"I think I see a scoter out there."

"Yes, and I see a zebra-striped elephant," Dan Scurlock agreed.

"No fooling," Rose Ann Rowlett insisted, and so our search began.

We were at Austin's Hornsby Bend sewage-evaporation lagoons, variously christened Petunia Ponds, Lily Lakes, Rose Acres, and other inappropriate names by birders. We estimated that this afternoon at least two thousand ducks were making a living here on algae, macroinvertebrates, duckweed, and other goodies nurtured by the biosolids that had settled on the bottom, and we had just set ourselves the task of singling out one duck.

As usual, the ducks weren't easing matters, drifting away as soon as we showed ourselves, so we retreated below the twelve-foot-high levee to shift positions several hundred yards, inching up till our eyeballs cleared. The levee's extrasharp rocks had scarcely dug in before one of us made a false move and the ducks were in flight.

"There it goes to the far corner," Rose Ann announced. How she managed to pick an individual out of the clutter was a skill I fervently hoped I could also achieve someday.

"I thought a heavy duck headed across the tank," Dan said, "but maybe not. Come on—let's get down to that corner."

"Uh-oh," I thought, realizing we'd be downwind. Well, any sacrifice for birds.

Only a quarter of a mile, and of course the youngsters sprinted it. Extra-extra-cautious and wishing for clothespins on our noses, we eased up on hands and knees again, and there was our bird, preening, definitely a pretty female scoter with the white spots before and after her eyes and a knobby bill, but no visible white on the wings.

"If she doesn't have wing patches, this is going to be one for the books," I bet. "Let's see those wings!"

Rose Ann Rowlett, professional field guide, in 1982. (Courtesy of Suzanne Winckler)

Of course, now our highly skittish bird refused to fly. Even when a birder named Harold arrived, moved in, and began waving his arms, our lady duck gave the merest flutter and swam away.

"I couldn't see a bit of white," Harold reported.

A Surf Scoter, a sea duck whose natural habitat is the ocean, here in Austin, Texas, a full two hundred miles from the Gulf of Mexico? Even in the Gulf this bird would be very rare. We had to make sure, but by now our duck was lost among the hundreds, right back where we had started. Panting, we sneaked up the levee again, and this time our duck showed white wing patches to all of us.

"So we know we have a White-winged Scoter, but Harold said he couldn't see any white. Do ducks play games with birders?" I wondered to myself.

Several more tries did nothing but leave us "quacked out." We could only pass word to three birders coming in, but they had no better luck with their telescope.

The levee around the 165-acres of man-made lakes measures a little more than two miles, and the next day I herded the duck swarm by car while three new recruits observed from the other side. Only a glimpse, but now we were sure there were two scoters, both females.

Next day, gambling that the wanderers would still be there, Mary Anne McClendon and Iris McDermott stuck with the multitude of ducks for hours with a scope. Just before dark, Mary Anne saw both birds, one bathing, the other stretching, and the wings told the story.

The combined efforts of ten persons over three days had established the first report in history of a Surf Scoter in Travis County, Texas, and the second report ever for a White-winged Scoter.

*   *   *

Now take a deep breath of fresh air and fast-forward more than three decades. The status of the scoters remains very rare, but, lo, the changes that have occurred at Lily Lakes! By the 1980s the City of Austin Water and Wastewater Utility had begun using the facility to produce Dillo (short for "armadillo") Dirt, an environmentally safe fertilizer made from sludge and plant materials. And working with the University of Texas and Texas A&M University, the city built the Center for Environmental Research at Hornsby Bend—a building with a laboratory, an auditorium, and two classrooms—to support research projects. Sharing a common interest in issues of sustainability and urban growth, the city and the universities generated a number of programs through the years. The Hornsby Bend facility has now received more than two dozen awards for environmental excellence.

Moreover, since 1961 the Travis Audubon Society in Austin had been conducting monthly field trips to Lily Lakes, and for years birders had taken for granted that these settling lagoons in the migration route known as the Central Flyway would always be there as one of the best inland places to see shorebirds as well as ducks. The bonus was the more than three miles of riparian woodlands along the adjacent Colorado River, woodlands that offered excellent land birding. Besides all this, there were fields and meadows in the twelve hundred acres of the complex that served as homes for many other bird species.

On any given day, birders could be assured of fifty species of birds around the ponds, and the entire unit could yield as many as a hundred. Hornsby Bend was noted for hosting very rare visitors such as the Northern Jacana, the Ruff, the Curlew Sandpiper, the Purple Sandpiper, the Ringed Kingfisher, the Great Kiskadee, and the Fork-tailed Flycatcher. A weekly bird count was begun in 1993 by Ian Manners, Barbara Parmenter, and Robin Doughty (all professors at the University of Texas), and by 1997 they had totaled 118 species. Regularly, local birders took out-of-town birders to Hornsby Bend, especially during migration. And the facility had become an important part of Travis Audubon Society's Christmas Bird Count circle, an annual survey of early-winter birdlife.

But a crisis came in 1995. With the closure of nearby Bergstrom Air Force Base and the development of new ways of handling sewage waste, the Water and Wastewater Utility decided to shut down two of Lily Lake's huge lagoons. For me, a shutdown of any of the lagoons of Lily Lakes would have been a personal loss, for it was there back in the 1970s that, with Red's help with telescopes, I took busloads of economically disadvantaged elementary schoolchildren to give them their first experiences of field birding. Now it seemed that this prime birding spot might be lost.

John Kelly was president of the Travis Audubon Society when this threat developed. He and Melody Lytle of Travis Audubon worked with others — including Bill Stout from the Water and Wastewater Utility and Robin Doughty and Ina Manners of the University of Texas Geography Department — to form a Hornsby Bend Steering Committee. Their activities awakened birders to the issue, and they began besieging the Austin mayor and city council with calls. Soon the committee had won the support of such influential organizations as the Environmental Committee of the Austin Chamber of Commerce. Their efforts resulted in a resolution passed in 1996 by the Austin City Council that committed the city to a policy of no net loss of bird habitat at Hornsby Bend. The Lily Lakes were saved!

Travis Audubon Society also undertook a number of commitments. It agreed to assist in the development of educational programs and materials for Hornsby Bend, to provide field trip leaders and equipment for student groups, to seek additional funding, and to pursue collaboration and expertise in wildlife management. Given the international nature of migratory birds, the resolution included a commitment from Travis Audubon to seek collaboration with Latin American conservation organizations, especially those whose purview included wintering grounds of many of the birds found at

Hornsby. The resolution also provided that, should any proposed modifications to Hornsby Bend result in increased bird activity with a detrimental impact on the new Austin-Bergstrom International Airport, action would be taken to reduce or eliminate these impacts. Travis Audubon is continuing its work to fulfill these commitments.

In 1996, Kevin Anderson, a doctoral student in geography at the University of Texas, organized the EnviroMentors, a program involving University of Texas undergraduate students (and now called the Ecological Mentorship Program). They did a GIS (Geographic Information System) mapping of Hornsby Bend, set up a Web site for it, and even began planning a trails system. The EnviroMentors developed mentoring relationships between university students and secondary public school students, many of whom were at risk of dropping out of school. Nowadays the Ecological Mentorship Program cooperates with the Colorado River Watch Foundation.

Each year, children from the fourth, fifth, and sixth grades at Hornsby-Dunlap Elementary School participate in the Living Lab program at Hornsby. It was my privilege to be there one morning when sixty schoolchildren arrived for an adventure.

After a short indoctrination at the research center, the students began their explorations in the field. The volunteers from Travis Audubon and Capital Area Master Naturalists working with the children on this day logged a combined total of one hundred hours.

"How do you spell 'damsel'?" Billy asked one of the volunteers. Billy knelt, and placing his folded paper on his knee, he carefully recorded in capital letters that he had just seen his first damselfly.

"They are supposed to write down what they see," a volunteer named Pat explained.

Each child had been given a small plastic sack. "What do you have in yours?" I asked Miguel.

"I have a leaf from mistletoe and a leaf from a hackberry tree," Miguel said, "and this is a stem from johnsongrass with seeds on it."

The group ahead of us had collected a small sample of greenish water from the pond. Back at the lab they would examine it to see what life-forms they could find.

At the bird blind, a pounded-earth construction built of all native materials by EcoFair Texas and others, Travis Audubon executive director Rob Fergus had set up a telescope. It was pointed across the lake at an Osprey, busily eating a fish it had just caught.

As the kids took turns looking, Rob said, "Just think, that bird doesn't have a knife and fork and certainly no hands. Watch how he is tearing the

Volunteer John Kelly instructs children at Hornsby Bend. (Courtesy of Kevin Anderson)

fish into small pieces with his bill. How long do you think it will take him to eat it?"

Alex was bored until he saw something crawling below the lookout opening. It was a small millipede, and John Kelly coaxed Alex to scoop it gently into his hand. Alex proudly showed it to the other kids as it crawled from one of his hands to the other. John was pleased as he told me aside that Alex was an "at-risk" child.

"How many legs do you think it has?" John asked. The fast-moving arthropod offered a challenge. "It's called a millipede. That means it has a thousand feet. Does it really have that many?"

The kids looked closely. "No way," Mary declared.

Since children in the fourth grade return to Hornsby Bend when they're in the fifth and sixth grades, there are opportunities to check just how much of their outdoor experiences has stayed in their memories.

It is surprising how many other opportunities the old sewer lagoons now offer, and the participants, like Topsy, just keep growing.

The Austin Youth River Watch program, for example, trains and pays students to monitor water quality at sites around the city. And, annually since 1997, the Austin/Travis County Health and Human Services summer youth employees have spent ten weeks learning about and developing trails and facilities, which were funded by the U.S. Fish and Wildlife Service and the Texas Parks and Wildlife Department. These local high school students have built more than five miles of trails—and there are more to come.

The summer youth employees also helped develop a Blacklands Prairie wildflower site and created examples of different types of home composting bins. Information kiosks have been built at various points around the ponds. TreeFolks and Texas Native Nursery have enhanced the habitat at the ponds and throughout the riparian forest. A botanical survey was conducted by a group of University of Texas undergraduates, led by Elizabeth Welsh, for a natural resource management course.

In 1999, Rob Fergus established the Hornsby Bend Bird Observatory to provide regular shorebird monitoring, a hawk watch, and a passerine banding station. Rob says the total number of species sighted at Hornsby will surpass 360. The records show that many other rarities, such as our scoter records, have occurred here since birders first discovered this area in 1959.

So far, more than a dozen organizations, including the Heart of Texas Peace Corps Association, are participating in making the most of the once-scorned Lily Lakes. The last Saturday morning of every month is designated Ecological Literacy Day, with groups taking turns working on projects and learning about local ecology and history. At least twenty potential research topics for the future have been identified, including urban agriculture, prairie restoration, environmental/land ethics, rainwater harvesting, green building, on-site treatment systems, and restoration ecology. It seems that the possibilities are truly limitless.

Nowadays the almost odor-free Lily Lakes and environs are designated as an important bird area by Audubon Texas, Travis Audubon Society, and the Texas Parks and Wildlife Department. The international Partners in Flight has awarded the Hornsby area the highest level of recognition in the organization's Flight Star program.

All of this wonder and more now exist because during a Thanksgiving holiday on November 25, 1959, an avid young birder named Frank "Pancho" Oatman noticed large numbers of ducks flying across the Colorado River. Figuring there had to be a pond of some kind nearby, he was the first birder to discover the birding possibilities at the sewage facilities of Hornsby Bend. At that time the facilities were often called Platt's Ponds, after the family who lived there and managed the ponds.

Pancho found many ducks and among them the first Travis County records of Common Goldeneyes and a Bonaparte's Gull. He alerted Fred Webster, Edgar Kincaid, and John and Rose Ann Rowlett, who went out the next day and found two other Travis County firsts: a Dunlin and two Lapland Longspurs.

My story, which started with a duck more than forty years ago, has now turned into a swan of a story. I marvel at it all and sometimes even wonder if bird people could save the world.

CANOE IN THE WILDERNESS

This was the actual telephone conversation in Lantana, Florida, in August 1965:

"Howard Langridge? This is Marjie Adams from Texas. My husband and I were told you might know where to find the Everglade Kite."

"Mrs. Adams, are there just the two of you? Can you both paddle a canoe? . . . Well, I think we can go in the morning."

At dawn, after laying eyes on us for the first time, Howard squeezed us, his son, and a twenty-foot canoe into his station wagon. Twenty miles later we began purring down a canal of the Loxahatchee National Wildlife Refuge, the canoe bobbing behind our rented boat. Very soon we left every sign of civilization behind, and as I gazed at the wilderness of the Everglades stretching to infinity in every direction, it occurred to me that we had now entrusted ourselves entirely to a man we had just met.

"There were seventeen kites here last year, some of them nesting," our host shouted over the motor. "Nobody's spotted any this year, but we may luck out."

Seventeen—possibly the entire U. S. population of this vanishing bird now found in North America only in the wilds of the Everglades. Constantly on the lookout, we churned mile after mile through the brown-dyed water. Then unexpectedly our route was blocked by the bane of the Florida waterways, the delicately blooming blue water hyacinth.

Though Red is athletic, he is not a big man, and I usually weigh about 110 pounds. Portaging a sixteen-foot boat, its motor, and a twenty-foot canoe over the eight-foot canal bank was not in the plan. Fortunately, Howard's son was husky, and the four of us managed it, and, muddy and sweaty, we sped on again under the cloudless sky to the ten-mile mark.

Here, as planned, we took the canoe back into the canal, but finally even it could not be forced through, and we made another canoe portage over the canal bank and out into the trackless marsh. For three solid, hard miles we paddled, making two more portages back into the easier canal wherever possible.

We saw Black-crowned Night-Herons, Common (Great) Egrets, Turkey Vultures, a cardinal or two, snakes, darting fish, and an alligator, but nowhere the floppy flight of the Everglade Kite. Luck definitely is needed with a ratio of 17 birds to 145,000 acres of wilderness.

The return trip was a replay of paddling and portages for a total of twenty-six miles in seven hours. We now felt we had known Howard Langridge all our lives and considered him a hero. In thanks for his efforts, we promised him a Golden-cheeked Warbler at Roy Creek, Texas, sometime in the future.

All that effort and no kites, and things grew no easier as, along with other birding, we continued to question anybody from birders to rangers, fishermen, hunters, Chamber of Commerce ladies, boat captains, zoo people, and service station attendants for any information about rare kites.

Then we got to Fish-eating Creek, scanned the well-known names in the register of guide "Skipper" Stem, and plunked down $15, about a fourth of a month's grocery bill in 1965.

In Stem's high-powered boat we raced twelve miles out into vast Lake Okeechobee where we quietly worked our way into the thick marsh of Observation Shoals for about a mile—and there was that rarity, the Everglade Kite. It had been so easy, it was almost criminal.

This male and his mate, which we never saw, had returned to this isolated marsh only three weeks earlier, the first kites sighted on Okeechobee in more than two years. Slaty blue-black, with bright red eyes and face and a broad three-foot wing span, this bird was a joy to view from so nearby as he hawked across the marsh with his wide, square tail spreading in an elegant black-and-white fan extraordinarily responsive to the air currents.

Swooping into the vegetation, he came up with an apple snail two inches across. Alighting as easily on the slender reeds as most birds perch on a limb, the kite held the spiral shell in its talons—with long, curved claws adapted to this task—and waited patiently for the shell's occupant to venture out. Then, with a rapid jab of his extremely hooked beak, the bird pierced the snail's nerve center, paralyzing it for easy withdrawal.

This kite is one of the most specialized foragers in the world, a specialization involving a price, for when the marshes are drained, the snails die out, and with them the kites. Hunters too have taken their toll on this completely gentle bird, killing it not for food nor for decoration but simply because it is a hawk. Stem, who has probably observed the bird more than any other person, thinks they are doomed in America unless new blood can be imported from South American countries, where the kites are comparatively numerous. Understandably, he hates to see the Everglade Kite disappear, for it has helped him make a living for years. As for Red and me, we felt we had gotten our money's worth from Stem and thanked him.

Three days later we received a card from Howard Langridge, which stated simply: "You missed the kites in Loxahatchee by three miles."

* * *

It was April 2002, and I was on the telephone again, this time talking to Ken Hollinga about his detailed report of the American Birding Association's regional conference in Miami, Florida, the previous January.

"Ken," I said, "you reported in *Winging It* that a busload of birders watched a Snail Kite catch and eat an apple snail." I then posed this serious question: "Since when do they drive buses out into the Everglades?"

Ken laughed. "It does sound odd, just to tell it that way, doesn't it?" he agreed. "But, Marjorie, it's true. U.S. Highway 41, the Tamiami Trail, runs by the Shark Valley entrance to Everglades National Park. Yes, the birders rode by on the tour bus, and we stopped and got out to watch the birds. There are usually several Snail Kites in this area—and there they were."

I didn't get the answer to why several Snail Kites were usually there. Were they trained? Baited? Bribed? Planted?

Larry Manfredi of Homestead, Florida, who leads bird and nature tours, answered it simply this way: "It's close to the canal and close to the Miccosukee Restaurant."

My conclusion about this event: birding is a game with millions of possible plots, and perhaps Red and Marjorie became birders far too soon.

Bus tours and kites are not the only things changed. First, of course, is that this species, once called the Everglade Kite, is now called the Snail Kite. This makes sense for a bird that has one of the most selective diets of any bird species on earth—snails.

Back in 1969 the annual census tallied 96 Snail Kites, and in 1972 only 65, indicating that the species was declining. The U.S. Fish and Wildlife Service began tagging young nestlings as well as adults, so that more accurate records could be obtained. In 1994 the census was 996 birds, and in 1999 the count was 3,500. The census in 2003, however, recorded 1,610 birds. From these changes in Snail Kite populations, we can see the effects of better management of water levels: the Snail Kite can be saved for now in the United States.

Snail Kites are gregarious and will roost not only with their own species but also along with a number of other species. In good seasons the kites may nest and roost in loose groups and take advantage of the plenitude by mating as early as nine months of age. Furthermore, scientists have learned that prosperity may lead a male or female to desert a nest, leaving its partner to raise their young alone while the deserter begins another nest with a new mate, a behavior that is almost certain to increase population. And these birds have now been recorded to live as long as thirteen years.

But this beautiful specialty is still on the endangered list and coded N-2 by the American Birding Association, meaning that once a person gets into the habitat and range of the species, finding it may be relatively easy.

So once again, we old-timers (after having walked barefoot ten miles in the snow to get to the one-room schoolhouse, etc.) look back in amazement at the changes that time hath wrought. Indeed, it was only the getting there that was hard in "the good old days."

HE GAVE US FREEDOM

"Of course, all of us treasure our stories about him and from him," I mused, "and it's a comfort to share them."

I was reminiscing with Pete Dunne, director of the Cape May Bird Observatory in New Jersey, about our hero, Roger Tory Peterson.

Pete Dunne was one of the truly lucky ones. He was saying, "I think the thing I value most about Roger is that when we were in the field together, he felt free to be himself. When you are a public personality, you have to wear a public persona. It's a defense to protect yourself. When we were together, he could take it off.

"Roger and I liked each other very much. He was tantamount to a god to me, but I guess I got over that, and he opened the door to me. Noble Proctor and I were among the select few people Roger called his adopted sons."

As Pete Dunne noted, not many had such a close relationship with Peterson. My good fortune was simply having the privilege of interviewing Peterson three times and being with him briefly in the field. Looking back at my first visit with this awesomely gifted man—an inventor, artist, and hands-on naturalist who caused a revolution—I can't help feeling pity for myself, an unseasoned reporter, who, at that stage, was totally bird-witched but only self-taught in the basics of bird science.

Red and I were attending the National Audubon Society's 1966 convention in Sacramento, California, and Peterson was a noted participant. I wandered into the convention headquarters, mentioned I wrote for newspapers, and inquired about Peterson.

"I think my husband would be pleased to talk to you," Barbara Peterson suggested. "He's right over there."

Roger graciously agreed to a visit, but where?

I had made a novice's serious mistake. I had not set up a location ahead where we could visit. I asked someone at an office if there was such a spot nearby. The man very generously offered his own office.

When I mentioned that my hometown was Austin, Roger immediately asked if I knew L. Irby Davis. Yes, my mother was a Davis, and Irby and I often called each other "cousin."

"As you know, Irby is an authority on the birds of Mexico," Peterson said. "He has made many trips there to record birdsongs, always camping out in the wildest places, which sometimes could be dangerous. I took my camera and went with him back in the fifties."

"Cousin Irby is not well," I reported, "but he's still working, and he's in some kind of conflict with Cornell Laboratory of Ornithology to get the return of his written descriptions, dates, locations, and other notes that were the key to his bird recordings made in Mexico. This means so much to Irby, I worry it may affect his health."

Peterson's great esteem for Irby was obvious, and he was saddened to hear that his old friend's health was failing.

About this time the man who owned the office we were in began to pace in front of the door. There was nothing to do but give his office back. Thus my interview with the Great Man came to an end sooner than planned.

* * *

Nevertheless, I found in my files the following interview with Peterson in Sacramento, California, dated November 14, 1966.

"I suppose I'm a night man," Peterson told me. "I seem to wake up about ten in the evening, and I'm going strong until about four, when the motor runs down. Actually, I like to work right through till dawn. The first hour of the morning is the best one to see the world. Then I have breakfast with my family, and then I go to bed. That seems the proper way to live, but most of my friends don't think so."

(I have a penciled note here: He *had* to revise his habits somewhat in order to meet up with birds—not that many are nocturnal.)

"We don't take a Christmas Count every year around my home in Old Lyme, Connecticut," Peterson continued, "for I am in some other place like the Congo or the Antarctic, as I was last year, but this year we're going to do it again. The real secret in a small town without too many really sharp birders is to import some clever talent from cities nearby.

"Today there are ten times as many competent birders as there were ten years ago. We have three rather good strong cities within a half hour or an hour's drive, and we can pirate some of their best observers. Give them a good spaghetti dinner afterwards, which makes it more fun—sharing notes and having a social evening—providing you have any life left in you, having started out at dawn."

"There are always puns on birds' names," I commented, "but the remarks made back and forth, some of which border on insults, were surprising to me."

"In Austin, Texas?" Peterson asked. "The home of Fred Webster and Edgar Kincaid? Edgar is one of our very best humorists and satirists. And Edgar is one of the great watchmen concerning an observer's 'unsanitary' bird records. I think 'unsanitary' is a very good term to describe inaccurate sightings.

"Allan Cruickshank has bestowed the title of 'Reverend' on Edgar, and I think he prefaces Edgar's name with it even in the Christmas Count reports. Edgar is often very irreverent. Records have to pass Fred Webster too, don't they? He and Edgar are the fine-tooth combs in Texas. Edgar knows the Hawaiian Islands very well too, you know.

"The big holdup on my Mexican guide is the flower guide. Without too many interruptions we hope to get it out in 1968."

***

I am interrupting this interview here with a long letter I found clipped to it. Dated January 10, 1967, it is from L. Irby Davis:

Dear Cousin Marjorie:

As to Roger Peterson ever really doing any serious work on the Mexican guide, I cannot say. I only know that each fall for the past five or six years Roger has promised faithfully that just after the New Year he would get at it in a big way. As you know, I wrote the text for it, and he did one of the color plates soon after I delivered the text; then he got off on a trip to Spain that lasted several months and he never did get back to the Mexican plates.

Then about four years ago, the publishers jacked up Roger quite a bit, and he pitched in and did three more plates, then dropped everything to go to, I believe, the Galapagos. Three years ago Eddie Chalif said he was absolutely sure the book would be finished because Barbara had promised Roger would finish it! Now for two years she hasn't even mentioned it in their annual letter. It has been ten years now, and my text is completely out of date and would have to be rewritten. You can see the old manuscript anytime you come by the house.

The upshot was that Cousin Irby found himself a talented young artist by name of F. P. Bennett Jr., and Irby's *Field Guide to the Birds of Mexico and Central America* was published in 1972 with forty-eight color plates. Among those given credit for helping with the voice studies was Roger Tory Peterson. That same year, E. P. Edwards's revision of his book *Field Guide to Mexican Birds* was also published.

Peterson and Chalif's *Field Guide to Mexican Birds* was published in 1973. Peterson had told me in the above interview that not all of the plates would be in color, but when it came out, it also had as many plates as Davis's book, and all of them in color.

There are major differences in the two guides. Irby Davis had his own ideas about taxonomy, and his taxonomy never conformed with that of the American Ornithologists' Union. He had a particular fury about "all those lumpers" (taxonomists who lump taxonomic groups together), and since song was his specialty, he often differentiated species by song variation. He was also inclined to give full species designation to the races that could be distinguished in the field.

Edgar Kincaid had gone to Mexico with Irby and felt Irby's many years of work deserved recognition, so he probably was influential in getting Irby's Mexican guide published. As more scientific information is obtained, it is possible that many of Irby's taxonomic positions may be confirmed.

* * *

I return here to another section from Pete Dunne's talk with me.

"Peterson, because of his many obligations and duties didn't have the time to take the pulse of the birding community as well as I did," Dunne noted. "I was traveling more freely at that time. I was extraordinarily flattered that he would trust my judgment and opinions. He would say, 'I've been asked to do this. What do you think about it?' Roger simply didn't have the time to do it all."

* * *

And now let's go back to my visit with Peterson in Sacramento. He was saying, "I need to do my eastern guide over now. I flinch when I look at it—both the western and the Texas guides are so much better—so I need to bring it up to standard. I'll be back in Texas during this work interval, but I won't be lecturing. In order to get the Mexican guide done, I must say no to as many things as possible. But there are some things, such as this Audubon meeting, that I can't say no to. I always give my full support here.

"My sons are age seventeen and twenty. Boys are not usually interested in what fathers are interested in, I think this is just male perverseness. The youngest one might become a naturalist. They don't accompany us on the trips at this time.

Roger Tory Peterson with teenager Kenn Kaufman, ABA Champion Birder of the Year, 1972.

"Advice to a novice? This is difficult—I'm so far from my early days with birds, some forty-odd years. My own interest, of course, as a boy of eleven was emotional—nothing intellectual about it. Birds are exciting, they fly, go where they want to when they want to. We'd like to have this freedom for ourselves. As for girls, I don't know, of course."

"I think beauty is what captured me," I told him. "Their extraordinary designs and combinations and their unbelievable diversity."

Peterson responded, "I think for boys it's more intangible than that. Birds are exciting, and boys don't question why. Birds reflect the life sources about as sensitively as any creatures. It's this life vitality. No bird without this strong life force will survive. Of course, as we get older, we begin to philosophize and intellectualize, and sometimes we lose sight of the emotional appeal."

"What about birding integrity?" I asked.

"A man can pull the wool over the eyes of people for a while," Peterson replied, "but integrity always eventually shows through. Here in California there is a wonderful lad from Scotland. [He was speaking of Guy Mackaskie, whom I interviewed some weeks later at midnight, the only time between work and birding that this devoted field man had to spare.] This young man is terribly good, extraordinarily good, and he lives for his birdwatching. He has come up with things too 'fantastic' for California, but he has always been able to prove them. Now he's no longer questioned.

"I don't collect, though I have a federal permit to do so," Peterson continued. "I haven't found it necessary. I have picked up birds I found dead and used them in my drawing, and I get skins from museums. I also take movies, and I have gotten some of my finds that way. Collecting done within limits very carefully by a few people, I don't object to. I don't think a man like Kincaid would do it—I know he wouldn't."

"However," I pointed out, "in the case of the collection of the Black-whiskered Vireo in Houston, Edgar did remark that the bird was a stray, and the collection did verify the occurrence. Of course, the great furor was from birders who were cheated of a Lifer."

"I think I would agree with Edgar," Peterson said, "because an accidental, a stray, is not going to survive anyway. For example, a European bird that turned up here will never find its way home, and there is no conservation issue involved. However, if it's a bird that might be a prospector and a pioneer for a spread in territory, then it should not be taken.

"As far as illegal collectors go, I know that Texas has a particular problem because there are so many unusual things restricted to a small area along the Rio Grande. California and Florida have the same problem. My friend Irby is completely against any collecting, and I agree with him most of the time."

Next I asked him, "In your travels have you found a favorite place that is a heaven on earth?"

"I have had at times favorite places," Peterson recalled, "but they all have their flaws or drawbacks. At one time I thought living around the Monterey Peninsula in California would be ideal, but so many people had the same idea that I'd rather stay in Connecticut right now. Then there's Africa. Much of Africa is terrific, particularly Kenya. I can't think of any spot in the world that would be more wonderful than parts of Tanzania close to Kilimanjaro, but the political situation would bother me. Yet for a naturalist and for the joy of living I can't think of another place quite like it.

"New Zealand is grand, but doesn't have quite enough birds to suit me. And part of it has been more man-modified than any other country in the world. It does have the largest national park to compensate for this. It is often we see this situation—man at his best and his worst. The genetic pool of man is so large now that we have everything from Genghis Khan and Nero all the way up to Christ and Buddha—all within our potential.

"I don't think I fit a conventional religious pattern, but I suppose I am a religious man. I don't know. I simply do not fit the conventional pattern and certainly not the formalized beliefs of any conventional church. I have been influenced by our Christian environment, but no formal religion fits my philosophy."

"Well, we're here with Audubon today," I said, "and I find most Auduboners open-minded."

"Yes, but you'll find even in a group like this a vast spectrum of ideas, and their approaches may be quite different," Peterson pointed out.

"Are you pessimistic about the future?" I asked.

"I'm a pessimist by nature. It is true we have a great proliferation of people, and the pressures are enormous and increasing. On the other hand, man is a check on himself. He should be able to find ways of preserving at least a part of natural life. This is why we have organizations like the Audubon Society.

"Man is still a young creature so far as intelligence goes. And he is an animal, not a god, but man does have a conscience, which does stir, and if we compare man today with man of three or four thousand years ago, this is a very short span in history. But things go so rapidly now, we wonder if man's conscience and intelligence can evolve fast enough to keep pace. This is what we worry about. We are still evolving, and we must hope."

**

It was a historic meeting, that June 1973, the very first convention of the American Birding Association held in Kenmare, North Dakota, and the man we all owed so much to was to be our featured honoree and speaker.

I was riding in the motor home, which regularly followed the seven buses that went on all the field trips. Roger Tory Peterson, who had slept late, arrived just as I stepped out of the RV with my binoculars in hand.

He smiled and nodded as he held up a cautionary finger. "Do you hear that Baird's Sparrow?" he asked by way of greeting.

I didn't.

He pointed, I cocked my ear, and I could hear it faintly in the far distance.

He laughed. "I'll tell you a secret. I have never encountered another person, except Chan Robbins, who has as good or better ear in the field than I do."

The sparrow was still singing. I realized suddenly that I was in an untenable situation. Remember, I am one of those sticklers who must *see* a bird to add it to my list. "But, fool woman," I told myself, "this is Roger Tory Peterson telling you in person that it is a Baird's Sparrow. Surely you can count this one." Oh, what fun the honest birding game can be!

About 150 people attended this first ABA convention. The good ladies of the little town of Kenmare, North Dakota, cooked our meals and served them in the high school gymnasium, our main meeting room. It was here that our banquet was to be served that evening, and it was here I encountered Roger Tory Peterson as he was just finishing hanging seven or eight of his paintings on the wall.

Of course I stopped to look at them, and the artist himself lingered by my side. "Oh, thank you, Kind Fate," I whispered to my inner self, "thank you that I was an art major in college. Oh, thank you indeed, that I have at least a somewhat trained eye, and I can look at these works in the context of art as in Art, the Great Experience."

All of the paintings were of birds, detailed realistically in their natural settings. I studied them carefully, one at a time, with the master quite attentive to my reactions. The pressure was great but gradually abated. The beautiful paintings should have pleased any nature lover as well as any aficionado of art. All had strong compositions, all handled color with sensitive intuition, and each made a statement.

The creative person is most naked with his or her work open to inspection. I began to feel that my opinion, humble as it was, meant something to the man at my side. Somehow, as we viewed his work, the field guides raced through my brain in a sort of weird comparison. The guides are a marvelous invention, a distillation, a tool keyed to the living bird that no painting can equal. Yet the guides are just that, a product dictated by need and use.

The paintings could be whatever the mind, heart, and soul of the artist could distill from his talent. I was looking at the real Roger Tory Peterson. Gone was the inventor, the field guide, the lecturer, the man with the best ear, the photographer, the teacher. The man I was talking to was solely the artist, looking at his work with a member of his public.

Was I wrong that day in Kenmare, North Dakota, to conclude that this was the way Roger Tory Peterson, the great bird artist, wanted to be remembered?

Through the ensuing years Peterson tirelessly sharpened his skills and techniques. The last of his works that I saw were displayed in the King Ranch Museum in Kingsville, Texas, as a part of the American Birding Association's meeting in April 1998 in McAllen. Those drawings were superb masterpieces. Without doubt they gave him the right to be remembered as one of the greatest bird artists.

* * *

Here I am today in Austin, Texas, more than a quarter of a century later, sitting at my computer. I turn on a tape recorder, and a shivering thrill rushes over me as the smoothly modulated and clear voice that lectured to thousands of nature lovers across the nation is speaking to Marjorie Adams just as it did in Kenmare in 1973. Following are additional parts of my interview that day with Roger Tory Peterson.

"When was the first time you ever heard or used the term 'birder' or 'birding'?" I asked.

"It seemed to be spontaneous. I've always promoted the term," he asserted. " 'Ornithologist' seems to assume too much, in the same way as 'artist' assumes too much. I think 'illustrator' or 'painter' is better. I like the term 'birdwatcher.' It covers everything from the ornithologist to the lady who watches birds from a window or the sportsman or the scientist. It is a rather broad thing, perhaps too broad, but it also in the eyes of the public denotes the image of the little old lady in tennis shoes. You've heard that cliché, but of course today the lady may be wearing spike heels and doing a good job in conservation.

"I've always promoted the term birder. As for giving a one-sentence definition of the game, I guess I would get back to something peripheral. My friend James Fisher used to say—how did he put it?—that birdwatching can be a superstition, an art, a science, a hobby, or a bore, depending on the observer. Birding, as I think of it, is basically a game or a sport of identification, where you're covering the countryside rather intensively to see what's there and using your special skills. It's a recreation and a sport thing that can be pursued in a modest way or intensively, as a tournament. It can be done at any level.

"Birds are indicators of the environment, and with more people birding and going out to look at them, there is an increasing awareness of the environment.

"Birds have very high metabolism," Peterson continued, "and they react more quickly to changes in the environment. A person watching them on a regular basis can note changes in the birds that can mean something has gone out of whack. So birds are important in that way. They are important in many ways, but I believe that this growing awareness is because so many more people are going out to look at things."

"What do you see as the side benefits of birding?" I asked him.

"One sees the environmental impacts," he responded. "That's probably the most fundamental thing. Aesthetics are also very important. It's the emotional quality. I don't think you ever forget it. Birds can go where they want to, when they want to, with abandon. They actually aren't all that free, but they symbolize freedom.

"Then we become aware of the natural laws we all are subject to, but it is a kind of contradiction in a way. The bird seems to represent absolute freedom and abandon, yet it is as subject to natural laws as it can be. It's a thing that attracts a person, yet it's a thing that educates one to cause and effect. There are a lot of by-products. For one thing, birding is a way for some people to find a relationship with other people.

"Birding is also a very good focus point for travel. If you have a focus point, the other things fall into place. Rather than just travel for the sake of travel, you go with an objective, and that makes things more meaningful. But birds don't have to be a reason for travel."

"For me, another reward is association," I added. "Seeing a certain bird can remind us of a person, a place, an emotion, or a particular adventure—something that is a part of your life experience that the bird is connected with."

Then I admitted, "I've been wanting to ask you some questions that might be impossible to answer. For instance, out of all the birds you've seen, is there one you can pick out as the most memorable?"

"There are so many of them," Peterson said, "and some of them have been very amusing. The rarest bird I've seen? Well, let's put it this way. One of the rarest is the Ivory-billed Woodpecker in 1941. We had spent three days looking for them on this big tract in southern Louisiana and had given up. We had gotten to the car and were ready to drive away, and there they were not more than a hundred yards away!

"I've had some adventures that involved danger. I remember going out in a rowboat to some islands off the coast of Patagonia to see some albatrosses, and we were caught in a great gale, blowing offshore. Waves were about six feet high. There were three of us in the rowboat, and it took us six hours to row three miles. The fisherman who took us out had a set of oars, and the second set was divided between me and my friend. We were facing the fisherman, so we had to row in reverse. It is very difficult to row in reverse. The problem was we couldn't miss a stroke without getting the boat swamped. It was a big job. It was exhausting.

"You don't drown out there. You die of hypothermia because the water is so cold. Two men who were there about two months earlier died in about fifteen minutes when their boat capsized.

"So we do have adventures that can be dangerous. Those things happen. Some of these adventures have been amusing, but I guess one on a recent trip on the very large *Lindblad Explorer* in the Atlantic was rather macabre. We had a bird that was seven thousand miles out of its range, and if you have a bird that is even a hundred miles out of its range, it becomes a rare thing and all the local birders want to see it. But here we had a bird that was in the wrong *hemisphere*.

"We had finished our tour of the Antarctic bases and were headed back towards Cape Horn. We had gotten near the Diego Ramirez Islands, which are way out in the middle of nowhere, and we paused to look at the albatrosses nesting on the cliffs there. Then we headed on. About thirty-five miles northeast I saw a tern that didn't look right—it certainly didn't look like an Antarctic Tern. It was struggling, and it finally got to the ship and landed in a lifeboat on the next deck.

Peterson gently corrected me here that a boat is what can be carried on a ship. Then he continued with his story: "So I climbed up the ladder and saw the bird's wing and tail feathers protruding, so I reached over into the lifeboat and grabbed it.

"I thought, a Sooty Tern, a tropical species, which would be far out of its territory, but it was even better—a Bridled Tern, for which there has never been a record south of the equator. There was only one spot in South America where they had been seen to occur at all, and that was in the West Indies.

"So I came rushing down to the lounge with the tern in my hand, and the first person I saw was the ship's doctor. He was poised with a hypodermic needle in one hand, and I could hear a woman yelling, 'Help! Help!' Apparently this woman had been going through some sort of emotional or mental crisis and thought this was a good place to commit suicide dramatically by jumping into the sea. She had been subdued, and he was trying to give her a sedative.

"I was standing there with this bird in my hand—"

***

Horrors! The twenty-six-year-old tape of the interview suddenly acts its age. It refuses to go on, no matter how I coax it.

Quickly I call Victor Emanuel. Surely he has heard this story and can fill me in on it, but Vic is in the Antarctic. So is Greg Lasley.

Later, at ABA's thirtieth anniversary celebration in Tucson, Arizona, I ask Pete Dunne, Stuart Keith, and Arnold Small, but none of them can finish the story for me.

Perhaps someone else? Paul Green, ABA's executive director, thought it fitting to briefly outline Peterson's story on the loudspeaker to the more than six hundred people present for dinner. When he got to the part where the tape recorder went silent, there was a shocked "Oh, no!" from the audience. Paul then asked anyone there who knew the remainder of the story to contact me. No one has.

But I have another option. The tape does pick up again briefly, with Peterson saying: "This bird was in pretty bad shape, but I knew George Watson with the Smithsonian was doing a thing on the birds of the southern oceans. So I said, 'George, I've got a record for you, a Bridled Tern, and I can prove it.'" The tape goes silent again.

None of my inquiries at the Smithsonian or elsewhere reveal any records of this bird, so the mystery persists to this page.

Be a hero. If you know the rest of this wonderful story, pass it along!

TOP-OF-THE-WORLD BIRD

By combining the sports of birding and mountain jeeping, Red and I experienced a day still well remembered more than forty years later. Our hosts were Arlene and Homer Reid of Telluride, Colorado, an old mining town nestled in a valley of the Rocky Mountains. It is adorned with a 365-foot cascading waterfall.

We four had three objectives: Blue Grouse, White-tailed Ptarmigan, and old bottles, but little did we Texans reckon where the last lay.

A wooden bridge across the swift and icy San Miguel River led us onto a trail fittingly marked "ROUGH—Jeeps Only" and into scenery probably equaled only in Switzerland, for within twenty-five miles there are fourteen peaks reaching altitudes of fourteen thousand feet or more. Roads here usually follow streams, but we also went through golden aspen forests so thick that sunlight merely flickered through and over lush mountain meadows dotted with small beaver lakes. From one of these clearings we spied our fourth Golden Eagle in two days.

The close sun had us peeling off layers of clothing, which we donned again as we climbed. Red spotted the Blue Grouse, and four birds cooperatively froze on the ground long enough to be filmed in 16 mm color.

One down, two to go, as our road grew progressively narrower and steeper until, novices that we were, we felt that every turn would be our stopping point. Not so, even when we labored up to Silver Pick Mine at 11,500 feet, where harebells and gentians bloomed among strange mountain ferns. Here we met a hospitable mining crew living in a shack of the last century. Though these rugged men called the Gray Jays that lit on their table "camp robbers," the birds were their pets. We were strangers to the feisty birds, and they refused to pose for Red.

Above the mining camp the conifers were more stunted; then, as we climbed higher, they were replaced by bushy arctic willows, a favorite ptarmigan food. However, there was no lingering for us here, and soon even these foot-high trees were left behind, with only scattered grasses and flowers to break the barrenness and monotony of the all-encompassing and ominous rock-streams. These veritable rivers of crumbling stone flow from the mountain summits in sporadic events, large ones capable of obliterating everything beneath them, including jeep trails and things upon them.

The rocky scratch in the mountain's side we now crept along was so narrow, there was not a tire width to spare at the drop-off, and it often was composed of freshly fallen rock or was oozy from mountain seeps. The arctic-alpine home of the White-tailed Ptarmigan was indeed a forbidding land and provided an excellent opportunity to concentrate solely on these chickenlike birds as we kept climbing.

We could hear the squeals of the pikas, locally called rock rabbits, and occasionally we saw one's harvest of the sparse grasses it was carefully drying into hay for winter food.

Our hairbreadth rocky trail now widened into a small promontory jutting above a sea of crumbling rock, and what seemed half of the state, including mountains in Utah, spread in an extraordinary panorama below. Incredibly, a deteriorating stone building here had been a boardinghouse in the old mining days, miles by mule from the grocer.

My pulse accelerated when it was disclosed that our climb was not over, and I was too excited to savor the lunch we spread here. Again we began creeping the rocky tightrope up the mountainside, with turns so sharp that even the stubby jeep took them at crazy angles and chugged around with crunching and sliding wheels. On one of them Homer had to back toward the 2,000-foot drop-off before, shaken but exhilarated, we halted on a shelf perhaps fifteen feet wide at 12,400 feet elevation. The red, yellow, and gray summit of our mountain, Wilson Peak, seemingly was within reach but actually was another 2,000 feet above us.

On a seventy-five-degree slope in the debris of another boardinghouse settlement, Arlene and Homer began to dig for yesterday's-trash-become-today's-treasure, while Red and I, giddy with altitude, devoured the view, somehow ran the frigid camera, then still dutifully scanned our glasses everywhere for the protectively colored ptarmigan.

From here the "road" was too rough for a comfortable footpath, but we tried to climb to the saddle so that we could continue, somewhat dazedly, to seek fabled birds. Forced to rest every other step to ease our pounding hearts, we actually made another four hundred feet.

Suddenly we were asking ourselves, "What is that coming from behind the mountain's saddle?" The apparition began approaching us and apparently was as startled as we were as we exchanged penetrating gazes. The hiker, unlike us, was not panting, and he stood staring at us, unbelieving, looking us over like the curiosity we were, and then he saw the jeep.

"Good gottamighty!" he exclaimed. He looked us over again, shaking his head slightly, and said, "I never thought I'd see a jeep up here." He laughed out long and loud as he headed down toward the valley. We didn't have sense enough to ask him about ptarmigans.

Having collected an assortment of twenty-six old bottles—ink, bitters, "Oel" (oil), spirit, and other types—and movies of the Blue Grouse, but not one ptarmigan cornered, we began descending the mountain in a sleet storm.

At the very first switchback, Homer had to back up the jeep three times. I found myself gasping apologetically, "Homer, I'm really, really sorry, but I just have to close my eyes."

"That's quite all right, Marj," Homer replied cheerfully, as he jerked the gears and churned the tires. "I always close mine too."

When we stood down in the valley again, we looked back and could see that the lofty peak above us was now completely white. We couldn't believe we had just been there.

* * *

The Adams pair owned not one snowshoe and, in fact, had never seen one. Now, with snow falling, we were realizing that our two-month dalliance with New Mexico and Colorado birders who had so generously taken us under their birding wings could cost us our chances with a Colorado prize, the White-tailed Ptarmigan.

Luck had failed us. We had crossed the Continental Divide three times, searching diligently each time we reached a minimum altitude of 9,000 feet; we had topped out as high as 12,400 feet on three spectacular jeep trips up (and down) steep, crumbling rock switchbacks; and unusually early snow had blocked us out of prime Mount Evans. Now, on October 3, falling snow was rapidly turning the landscape white in Rocky Mountain National Park. Eight-foot drifts had closed the park's famous Trail Ridge Road. But thirty minutes after our arrival, snowplows admitted us to a sky-high world of splendors almost stupefying to us Central Texans, who in our entire lifetimes had seen an accumulated total of perhaps two feet of this fluffy white stuff.

Dressed in everything we owned and further burdened by heavy tripod, long-lens camera, and other photographic equipment, we trudged the recommended Tundra Trail area, often plunging unexpectedly into snow above our knees, with nothing but the horizon between us and the arctic wind. In little more than an hour, nipped to the bone and exhausted from lack of oxygen in the twelve-thousand-foot altitude, we had to give up.

Early next morning we continued our search beyond the Rock Huts in slightly more sheltered but much rougher terrain that offered a broken leg at nearly every snowbank and, at times, a long, long descent if we erred.

True to the legends, we were right on top of seven beautifully camouflaged ptarmigan without seeing them, but our luck had changed—they flushed!

This time we had gone searching without the camera, so Red toiled back to get it. I stood for forty-five minutes, figuratively and almost literally frozen, keeping the flock in view as the birds expertly nipped shoots from the dwarf arctic willows rising above the windswept slopes until Red returned.

For more than an hour Red coaxed the frigid camera to turn. With filming finished, I dared to test the noted tameness of these "game" birds, which rely so completely on fool-the-eye for safety. Yes, I would have touched them with one more inch.

We Hill Country Texans had some lessons to learn. Not accustomed to arctic conditions, we ignorantly were not properly protected, and my face had come so near to freezing and was so swollen that I was unable to risk it outdoors again for five days—a painful reminder that the elements rule, and birding can be a dangerous sport.

RAREST GIVES WAY TO
MOST ELUSIVE

"Mrs. Adams, could you tell us the rarest birds in the world so our artist can illustrate your article with them?"

The caller was Susan Swetser, art researcher with the *Reader's Digest,* and the illustration would accompany my article in the *Digest*'s June 1975 issue, under the title "Birding—A Sport for All Seasons."

I stumbled along, naming a few endangered birds, before the truth dawned that not even the most brilliant ornithologist could answer this question accurately. Who really knows how many Ivory-billed Woodpeckers, Molokai Creepers, Bermuda Petrels, or Australian scrub-birds are still alive? Has anyone seen a Seychelles Owl or a New Zealand Takahe just lately?

In *James Fisher and Roger Tory Peterson's World of Birds,* the authors list 143 species that have populations of two thousand individuals or less—many of these species, alas, down to a handful or so.

The task of finding an appropriate illustration for my birding article began shifting to the question of how many birders are trying to find these rarest of all birds. Not many, so I suggested that the illustration include instead some of the most difficult birds to find.

Naturally, at this point I turned to Stuart Keith, Champion Birder of the World, who as of 1975 had seen and identified in the field more than half the 8,600 species then known. According to Keith's publications, some of the most difficult to find are the shy forest birds such as the Long-tailed Ground-Roller of Madagascar, the Congo Peacock, and the Great Argus Pheasant of Southeast Asia. For such birds as the Cinereous Tinamou, the Violaceous Quail Dove, or the Rufous Antpitta, you may slog through muddy forests inhabited by inch-long ants whose bite can send you to bed for two weeks, a forest also home to bushmasters, giant relatives of the rattlesnake.

There are 45 species of tinamous, a few which live in open country, but the remainder are hidden in deep forests or dense brush. Keith thought that just

Among the rarest birds in the world are: (1) hooded mountain toucan, Peru and Bolivia; (2) white-spotted crake, male, equatorial Africa; (3) twelve-wired bird of paradise, male, New Guinea; (4) blood pheasant, male, Tibet and southwest China; (5) banded pitta, Malaysia and Indonesia.

Reader's Digest, *June 1975, publishing a painting of hard-to-find birds by artist Guy Tudor.*

to find and see all of these secretive, small chickenlike birds in itself would be a lifetime task.

Have you searched a marsh for a Black or Yellow Rail? Then you understand what Keith was saying when he reminded us there are 132 species of rails, and the foreign ones are just as difficult to find as our rails are. There are also the elusive game birds that inhabit jungles, mountain forests, and deserts. What about the many look-alikes such as *Empidonax* flycatchers—not to mention nocturnal birds like the 134 owls and the 67 species of nightjars?

Fortunately, Ms. Swetser and her artist were not far from the American Museum of Natural History and Stuart Keith himself, so they could turn those resources for additional help as needed. It was interesting to see what the artist finally came up with: the Hooded Mountain Toucan of Peru and Bolivia; the White-spotted Crake, Africa; the Twelve-wired Bird of Paradise, New Guinea; the Blood Pheasant, Tibet and southwest China; and the Banded Pitta, Malaysia and Indonesia.

It was a very unusual situation, but when asked to locate the painting so that it could be reproduced here, the *Reader's Digest* staff couldn't locate it or any records of it, and the search extended off and on for several months. They did come up with the name of the artist—Guy Tudor—and they thought he was in New York.

The excellent research staff available to the public by telephone at the John Henry Faulk Central Library in Austin searched the directories for New York State and came up with two persons named Guy Tudor. The first one I called was in Forest Hills, New York, and he was indeed an artist, and, yes, he had done some work for the *Reader's Digest* in years past.

"What is the painting like?" he asked. After I described it to him, he said, "I don't remember it. I must have sold it long ago."

So here the painting has been reproduced as well as is technically possible from a page torn from the artist's own copy of *Reader's Digest*, with best wishes from the artist to you, the reader.

THE TEXAS BIRD LADY

The neat white cottage was surrounded by flowers, and on the wall of the veranda a sign announced the hours allowed to visitors.

The door was opened by a petite woman, queenly rather than pretty, and gracious as we introduced ourselves. She agreed to be recorded as she shared some of her photos, bird objects, and other memorabilia with us. I especially remember a perky wren made of clear blue crystal.

Here in her own words is what she shared with Red and me during our visit on April 25, 1966, in Rockport, Texas, with Conger Neblett Hagar, known on two continents as the Texas Bird Lady.

* * *

"Yes, I did do a Big Day on April 25, 1953, with Clarence C. Brown of Montclair, New Jersey," she said. "He was staying here at the cottages at the time, an amateur like myself. He was a member of the exclusive Urne Bird Club up there, one of those that don't allow women at all. They told me that if I would come and lecture to them, they'd be glad to have me come in. I told them, 'Excuse me, if the other ladies can't come along, then I won't come either.'

"A Big Day had been suggested to me many times, but I had hesitated on a Big Day because I felt that on things like that you may try to see too quickly and call too quickly on your identification, so I hadn't been in favor of it, but Mr. Brown, an ornithologist of good standing among the clubs and professions up East, had done it several times with the Urne group, and he was very anxious to try and compare it to what he could do here.

"Yes, we did start out with a Big Day in mind—that was our objective. It was the first attempt in this area.

"Our count started at four AM. and lasted until eight PM. We took an hour out to come home for lunch. We had a perfect spring day, everything just perfect. Of course, we had breakfast, then began right here at the cottages with a couple of owls. We got the others later, just this side of Tivoli.

"I was using Zeiss 8-by-40s, and we didn't use a scope. We didn't have to. We traveled in the car almost continually, unless we wanted to count numbers, because there is less disturbance of the birds. If you get out, they all fly away, and that's why I don't like to bird with a crowd anymore. Some fool is going to get out of the car.

"When we get one of those big pushes of cold air coming through, that's when we get so many warblers. We had a norther the day before, and that is what held them down. Pauline James held her class out on Padre Island that day, and that's how I knew the warblers had stopped there. That's what gave us the numbers.

"Everywhere I go in my territory, I run into water. Out to Rattlesnake Point five miles, and I hit water. To the end of the causeway where the hummers start, and I'm in water. Go out the other way to the cover, and I'm in water. Not a big territory here—about seven miles long. I don't know how many square miles. The count was all in this area.

"We both kept a list, and then we checked them out together when we came in. We had 204 species. [In one day on a very small part of planet Earth they had totted up as many species as most birders can tally in a year in an area the same size.]

"One of the unusual birds we saw was the Glaucous Gull. They took my word for these things, because all the authorities from the government down —Dr. Harry Oberholser, then [Ludlow] Griscom from Harvard, [S. Dillon] Ripley from Princeton, Dr. Aldridge from Cleveland—all these authorities came and went with me in the field, and until then no one accepted a thing I said.

"Would you like for me to tell you something Dr. Oberholser said to me? He wrote from Washington saying the government was interested in lists he heard I made every day, and he would be glad to have a list. I told my husband, well, I'll send him the list I made today. And that's where fools rush in where angels fear to tread. In I went. I got this letter back: 'Dear Mrs. Hagar: We have received your list, for which I thank you. On this list we notice a Sooty Shearwater. How many days did you see a Sooty Shearwater in Rockport, Texas? What was the Sooty Shearwater doing? What books have you used for identification?'

"Right there, my little Yankee husband blew up. He said, 'Throw that damned thing in the wastebasket. No man can contradict my wife's words! You're not going to answer that damned thing.'

"I said, 'Listen, little boy, if I didn't see a Sooty Shearwater in Rockport, Texas, then they should identify it and tell me what I did see, and if I did see a Sooty Shearwater in Rockport, Texas, the government should know about it.'

"So I wrote down all the books I used, the names of the five people who saw it with me, and Terry Gill had taken a picture with one of those old Kodak-type cameras, and I put that in the letter too. I told Oberholser, 'I think if you put your magnifying glass on it, you can identify it, and tell me what I've seen.'

"Oberholser wrote back: 'Having the picture is the next best to having the bird in the hand. You did see a Sooty Shearwater in Rockport, Texas. You have taken a Pacific, oceangoing bird and placed it on the quiet bays of Rockport. You have made a record for your state.'

"So I found that was the only way to do it.

"When he asked me about a Warbling Vireo nesting in my own backyard, I wrote back that I had climbed a stepladder and looked at it. He said, 'You've taken a bird and put it just four hundred miles out of its nesting range.'

"I went through the same procedure with Roger Tory Peterson. He came and went through several migrations with me, and he said, 'You have me all confused.' I had said several times that warblers came and walked right over my feet in my yard. Well, when they walked around his feet, he liked to have died.

"This is when a norther hits when they are migrating north or when it catches them when they are going south, and they are very tired and drop down and start feeding. The Bay-breasted and the Hooded are two of the commonest to do this. The easterners thought they didn't do it, but they did.

"One time Roger, Guy Emerson, and Edward Chalif were all here to go through a migration with me, and we had a good one. They doubted me about Buff-breasted Sandpipers. So I said, 'See there, what's that in the field?' Peterson said, 'Well, they're Buff-breasted Sandpipers.'

"So Guy Emerson got out, eyed the barbed-wire fence, then lay down on the wet earth and rolled under it. Chalif, who is a ballet dancer, put his hand on the post and bounced over the fence. Roger pulled the wires apart, and when he eased through, he tore his pants. I sat in the car, watching the three monkeys perform. When they came back, they kept saying, 'I am amazed. I am amazed.'

"Well, Rockport is the only place I know that Buff-breasteds do this way and build up, except in Argentina when they are building up to come this way, so I've had hundreds and hundreds of people, including the authorities, to come and see if I knew what I saw.

"As Griscom told me, when I first reported them up there, he said, 'You don't mean you see them every spring.' And I said, 'Yes, every spring.' And he said, 'Now, maybe one or two.' And I said, 'No, I see them in groups, in numbers at times.' He said, 'Mrs. Hagar, I've been in the field thirty years, and I've seen two, and I'm the envy of the eastern ornithologists.'

"I said, 'Mr. Griscom, if you'll come when I say 'Come,' I'll show you some. So he flew down, and I took him out and I said, 'What's that?' And he said, 'I'm getting out of this car,' and he worked his way around. When the sandpipers come in, they start feeding, and they don't mind your walking by them. He came back, and said, 'You know, those were Buff-breasted Sandpipers.'

"When I reported on the migration of hummingbirds, I watched it five years before I reported it, and this after I had been reporting for fifteen years. You should have seen the people come. Edgar Kincaid was one of them, and Mrs. Dobie came with him. The birds swarmed all over them.

"Another time I sat in my backyard all one day, except to come in to lunch, and stayed all afternoon, and I got eighty-six species!

"Peterson's western book needed to be revised so badly, and a man from California came in sputtering, saying that if Peterson didn't finish the western book before the Texas book, it would be too bad when Peterson came to lecture there.

"I said, 'Who's paying for the California book?' I said, 'The Texas Game and Fish Department has paid thirty-seven thousand dollars to have the Texas book published. So that's why he did the Texas book first.'

"I banded birds for eleven years for the [U.S.] Biological Survey in Corsicana before I moved to Rockport. I organized a nature study club because when I was a little tyke six years old, my father took me and my sister, who was four, by the hand and led us out under the trees, and he said, 'Now you two little girls are Texans, and I want you to learn everything beautiful in your state. All the butterflies, birds, trees, flowers.'

"They had to get us through college pushing and pulling. My major was music, and my minor was English. They didn't teach any such thing as ornithology then. When I got up into my thirties I said, 'I don't play nor sing as well as I did when I was younger, so I'm going back to nature, like my father taught me.' In the nature club someone suggested I do banding to see what that was like. We trapped then. Government showed you how to build your trap. First one was, just prop it up and pull the string out when the bird came in. Then they improved it to where you pushed the door up, and it set the trigger inside, and the bird would hit the trigger and that would shut the door.

"As far as birding in your own backyard, you can quote Bob Allen. The ambition of his life was to do for Tavernier, Florida, what I had done for Rockport. Of course, he died before he could do it.

"I think it is a duty to let others know. Others then have a key, which they can work. I think the intellect is the only thing that's going to last anyhow. Don't get me started on thought—we're all living off of other people's thoughts.

"I think about birds just like Roger Peterson. I don't think you can put personal feelings in birds. I think it is all instinct. I think the basic things born in a bird are food and mating. I think everything they do relates to one or the other.

"Sometimes I think birds are a good deal like men. Food has a lot to do with how they behave."

***

From a visit with Connie in September 1965, I carried away the following memory.

"The first thing to do is to prepare your mind," Connie said. "When your mental attitude is prepared, you are ready to see things you have never seen before—rocks, flowers, stars . . . birds.

"I remember a lady from Yoakum, a beginner, who went far away to see a Painted Bunting. She didn't find it, but when she came home, she saw it in her own backyard.

"I was showing a professor a merganser, and he began to look at other birds. I began fussing at him: 'Look at the merganser. Really look at it.'

"And I don't believe in talking while you're looking. Durn that kind of looking!"

***

L. D. Nuckles, regional information officer with the Texas Parks and Wildlife Department, told me, "I never saw Connie Hagar mad about anything. She had no business driving when she got up in years, and she backed up and scattered a flowerpot all over. She sometimes bragged that she could cuss, but she didn't lose her temper.

"A dignified old man brought a bird to me that I immediately knew had to be a Jacana, but it was the wrong color. I took it to Connie, and she said, 'No, it's not the wrong color. It's just right for the second-year bird.'"

It was Nuckles who wrote the resolution naming the Connie Hagar Wildlife Sanctuary in Rockport, and it was Roger Peterson who unveiled the first sign on the Great Texas Coastal Birding Trail, dedicated at the Connie Hagar Cottage Sanctuary in Rockport on September 8, 1995.

On one of my visits with Roy Bedichek, he told me he called Connie when he was in Rockport. He wanted to meet her, and he asked her what she had been doing.

"I've been looking at a Least Grebe's nest," Connie told him.

"What!" Roy exclaimed. "I've never seen a Least Grebe."

She said she was surprised that anyone as elegant as he was would want to watch birds with her. She wore a dress over a bathing suit and rubber boots. She always said she believed, "Don't go unless you're prepared, and be ready when you get there."

Roy was ecstatic to see the little grebes going in and out of their nest.

As Roy and Connie traveled farther, he saw a bird and asked, "What's that?"

"A Piping Plover."

"You don't mean that's a Piping Plover," Roy said. "I've been all over the [Rio Grande] Valley and out with several people, and no one showed me a Piping Plover."

When it came time for them to part, he confessed, "Do you know why I came down here and wanted to meet you? I came down to prove you were a nature faker!"

After this wonderful visit, he fittingly called the tiny bird lover "Least Grebe." He was famous for his letters, and in one to her he wrote, "Dear Least Grebe, I am so pleased to know anyone who is able to get in the *Christian Science Monitor.*"

Connie was a speaker at the local chamber of commerce, and when a gentleman passed her a drink, she declined, saying she drank only a little beer at home and never drank out. Another gentleman offered a cigarette, which she declined, saying she never smoked. She declined coffee, saying she didn't drink it at night.

"Come now, Mrs. Hagar, don't you have any vices?" the man asked.

"Yes," she answered, "I can cuss like a sailor. Do you want to hear me?"

I met Olga Clarke of California when she and I were the only women on the American Birding Association's first organizing board of directors. She had been introduced to birding by her future husband, Herbert, and in fact Herbert had done quite a bit of birding on their honeymoon.

"We went to Rockport, and of course I met Connie Hagar," Olga told me. "I complained to her that I simply didn't understand all this chasing around after birds."

"Connie said, 'Well, Olga, he could be chasing a lot worse things.' I thought, 'Well, she has a point,' and I began chasing birds too. Now I've chased them on three continents."

Connie was a "locality birder," patrolling the same territory for thirty-five years. This kind of birder contributes the most to our knowledge of birds with year-in, year-out studies of behavior, nesting, migration, numbers, feeding habits, and voice that can be accumulated only through long hours of fieldwork. Connie added at least twenty-five species to the Texas state list.

She was a five-foot, hundred-pound bundle of energy, wit, and wisdom who graciously overlooked the ignorance of both novice and professional to share with complete generosity the store of knowledge she had so painstakingly gathered.

In my files I find a note from her written in 1963, which ends with "Always enjoy meeting some one who can t-h-i-n-k." Signed, "Least Grebe."

A treasure to be kept forever.

HUNTING — A BATTLE CRY?

"It's hard for me to understand how in the world you can kill such a beautiful animal," one of the guests at the party challenged my husband, Red.

"That's because you haven't spent as much time in the woods as I have," Red replied. "If you saw a deer suffering from starvation and tortured by parasites the way I've seen 'em—or if you saw all the deer carcasses of a die-off the way I have—you'd realize that a fast death from a hunter's bullet can be very humane.

"It's natural law," Red continued, "and we city folks aren't reminded in our daily lives how the natural world works. It's basic: living things live because other living things die—from the wheat in your bread to an oyster. And now we have so many people on earth that what you and I do every day affects other living things, from bugs on up to deer.

"Let me ask you, have you seen a mountain lion, a wolf, or a coyote lately? They're the ones that feed on deer, but they're all gone around here. So we have a lot more deer. And what happens? The deer strip the land—and then not only a whole lot of deer but almost all other wildlife are in competition for what's left to eat.

"So it's up to the hunters to keep the deer population down to what the land can safely sustain. It's just that simple."

"Can't we get some kind of birth control?" the guest asked.

Red laughed. "What taxpayer would be willing to foot the bill for the research and the number of people that would be involved? Not to mention the hullabaloo some religions might raise. No, just forget prevention.

"Deer aren't like cows. They don't eat much grass. What they like best is the new sprouts on shrubs and trees. They eat young wildflower plants before they've had a chance to bloom. You can tell where there are too many deer by looking at the landscape. Everything that's below how high a deer can reach—the browse line—has been eaten. That's a place where the forest hasn't had a chance to renew itself. And of course this hurts not only hungry

The thrill of hunting: Red Adams with deer harvested on Bonita Ranch, Texas, 1934.

deer but all the other forest creatures too. There are long scientific studies that have proved all this."

His listener was not convinced. "I contributed money to have deer trapped and shipped to Mexico," she said. "That's how much I hate hunting."

"Well"—and Red couldn't help a chuckle—"a lot of Mexicans are going to enjoy a fine venison dinner."

I will never forget a short hike Red and I had in Big Bend Park, a hike

The labor of hunting: Red Adams butchering a deer. (Courtesy of A. D. Stenger)

punctuated by the carcasses of fourteen deer, all sad proof that Mother Nature will control population density of a species, if not one way, then another. Is it appropriate here to mention that this includes humans?

Still, there are many who accuse hunters, wildlife "managers," and the munitions industry of engaging in a cruel and heartless game of death against the wildlife that belongs to all of us.

Thus hunting can be a complex subject. Many hunters, including Red, look forward to the hunting season each year not only for the challenge and camaraderie of hunting but with a feeling of obligation, of duty.

Bagging a deer usually involves considerable work and effort. Most often the hunt takes place in rugged country. The hunter must stay hidden or sit stone still in bleak weather. If a deer is wounded, the sportsman is duty-bound to pursue it to finish the job. The deer, a large animal, must be gutted, and then it must be carried from the field. It must be hoisted to hang for a day or two in a tree or from a scaffold, and then it must be skinned and butchered. This is all work—no other word for it.

There is much to be said for the hunter from a conservationist's view, because the fees paid by sportsmen often furnish most of the support for state fish and game departments and for refuges. It is a researched fact that wildlife areas bought with the aid of hunters' license fees support more nongame species of wild creatures than they do game species. Most birders appreciate

A bachelor party included some of Red's hunting and fishing buddies. Back row, left to right: Russell Tinsley, A. D. Stenger, Cactus Pryor, Red Adams, and Joe Small. Front row: Grady Allen, John Henry Faulk, Russell Lee, and Ken Hagen. (Courtesy of Jean Stenger)

this but chafe, nevertheless, when areas may be closed to the public during hunting season. Remember our old kindergarten rule: Take turns. Hunters' dollars count a lot.

Texas presents special problems because it has been estimated that about 20 percent of the total deer population of the entire United States is in Texas. At Roy Creek, Red and I had our share of deer neighbors. Mama and Greedy, a doe and her young fawn, came to us during a severe Texas drought. Red saw how thin the doe was and put out some corn. Thus the Adams deer herd began. In two years it numbered ten animals.

When hunting season approached, there was a debate in the household. It was that nitty-gritty question of the hunter's duty. Yes, it was true that at least one of "our" deer needed to be harvested, but which one? It certainly couldn't be Mama or Greedy. Nor Little Buck. And what about Feisty?

I'm reminded of a recent conversation with Jerry Cooke, director of the Big and Small Game Program at Texas Parks and Wildlife Department. Jerry told me something his dad had advised him: "If you plan to eat it, never name it."

When Red did shoot the selected deer in the head, it dropped silently to the ground, and, strangely, not one of the other deer raised its head or paid attention.

I must admit that corn-fed deer made some of the best venison I ever tasted, and it tasted even better based on duty performed conscientiously.

A couple of years later Red and I were sitting at the picnic table deep in Roy Creek Canyon. A deer approached, then stood gazing at us.

"Greedy?" Red asked.

The doe came a few steps closer. Red held out a piece of bread.

Greedy advanced another step or two, then stopped. Red threw the bread to her. She took it, turned and retreated.

As she disappeared, I clasped Red's hand. No other hunter's gun had taken her, and both of us let out a thankful sigh of relief.

As I mentioned, hunting can be a complex and controversial subject.

CRISIS IN GOOSELAND

While Red was hunting with other members of the Texas Outdoor Writers Association at the Blue Goose Hunting Club in Altair, Texas, Marjorie was exploring the Attwater Prairie Chicken National Wildlife Refuge nearby. On her return the writers needed photos to liven their articles, so she agreed to pose with some of the geese from the morning's hunt.

In the 1970s when this photo was taken, the bag limit for Snow Geese was two per day and four in possession. The Snow Goose population was estimated to be about 2 million. By the year 2001 the combined Snow and Ross's Goose population had grown to an estimated 5.8 million. Therefore, since 1980, the bag and possession limits have been raised, and as of this writing in 2005, during the regular hunting season the bag limit is twenty of these geese per day and the possession limit has been eliminated. However, much more goose control is still needed, so a Light Goose Conservation Order allows unlimited hunting of these two goose species for an additional period. However, the hunting harvest has been lower than expected because surviving older geese are warier, and a flock can have as many as twenty thousand pairs of eyes ready to note danger.

If you have been in Texas in winter in the past decade, you might have had Houstonites driving you out to coastal farmlands to witness as a recreation the vast hordes of Snow Geese feeding and resting in the fields. The bird throng turns the landscape white, and when the birds take flight, the sky becomes a snowstorm, with their combined honks making a clarion thunder.

Thus human success with farming and conversion of marshes to agriculture not only has helped humans but also has been hugely successful for wintering geese, which have discovered the superior food value of farm crops. This sometimes results in severe crop destruction.

The end result is that individual geese are living longer, and thousands of fat geese survive the hardships and rigors of migration and winter to return

(Courtesy of Texas Outdoor Writers Association)

to the tundra in the Arctic and subarctic Canada to nest. In the process the huge numbers of these birds are destroying the environment on which their lives depend.

More than thirty other migratory bird species such as Northern Shovelers, American Wigeons, Yellow Rails, Stilt Sandpipers, Short-billed Dowitchers, Red-necked Phalaropes, Lapland Longspurs, Semipalmated Sandpipers and

Hudsonian Godwits also nest and raise their young in this same fragile tundra. Because of the destruction caused by the multitudes of geese, these other species also face grave danger.

Snow Geese have the unusual feeding habit of using their serrated bills to grub out of the just-thawing earth whole roots and rhizomes of their favorite grasses and sedges. (The award-winning international film *Winged Migration* includes a scene of a Snow Goose in the act of pulling an entire plant out of the thawed tundra.) Ordinarily, enough rootlets would be left to revegetate the tundra, but the geese and their hosts of goslings eat everything bare. During their molt, a time when they can't fly, geese now will walk many miles to new territory to devastate it as well.

There are few predators of the Snow Goose on its breeding grounds. The cruel fact of nature is that the healthy adult Snow Geese are not going to stop nesting and laying eggs, and the tundra will be the scene of starving and dead goslings left behind when their parents fly south.

The damage to the breeding grounds occurs when the overgrazed tundra soil, left unprotected by vegetation, rises in temperature and thus allows evaporation, which in turn causes natural salts to accumulate at the surface. This high soil salinity prevents the usual vegetation from growing back, and the tundra soil itself may blow or wash away. Scientists now believe that without drastic change the damage could become permanent.

The startling conclusion is that to prevent irreversible damage, the current Snow Goose population needs to be rapidly reduced by half. Of course, this is an objective exactly the opposite of the decades-old practice of protecting and increasing waterfowl numbers.

In an effort to bring back some sort of balance, Native tribes have been allowed to harvest more eggs and adult birds, the hunting season has been extended, efforts have been increased to allow hunters access to private lands, and some refuges are allowing hunting. The situation has seemed desperate enough even for refuge managers to consider netting and trapping large numbers of geese.

After extensive consultation with the Canadian government and a rule-making process that generated hundreds of public comments, the U.S. Fish and Wildlife Service passed the Light Goose Conservation Order, whose rules apply after the regular waterfowl and crane hunting season closes. This special management action removes limits on daily and possession numbers of Snow and Ross's Geese, and guns are no longer limited to holding only three shells. Electronic calling has also been allowed.

The Fish and Wildlife Service's action was supported by the Canadian government and a broad spectrum of the conservation community, including the

National Audubon Society, the American Bird Conservancy, the Ornithological Council, and Ducks Unlimited.

With 44 percent of the Central Flyway's Snow Goose harvest occurring in Texas, the Texas Parks and Wildlife Department along with the Texas Association of Community Action Agencies, Inc., created a pilot project within the Hunters for the Hungry program to use Snow and Ross's Geese to help feed hungry Texans. Hunters can take their geese to a participating waterfowl processor and for a fee have the birds made ready for use by a food bank or other assistance provider, or hunters can donate their birds to a charitable organization.

It seemed that the Light Goose Conservation Order might solve the problem, for waterfowlers are a free, highly motivated workforce with an understanding of the problem and the objectives. "But it turned out to be more complicated than that," James R. Kelley Jr., wildlife biologist with the U.S. Fish and Wildlife Service at Fort Snelling, Minnesota, told me.

"On March 22, 1999, a lawsuit was filed in the U.S. District Court of the District of Columbia to stop the Conservation Order entirely." Kelley explained. "The plaintiffs in the U.S. lawsuit were the Humane Society of the United States, the Animal Alliance of Canada, the Animal Protection Institute, and the Canadian Environmental Defense Fund. A similar lawsuit was brought against the Canadian government by the Animal Alliance of Canada, the Animal Protection Institute, the Canadian Environmental Defense Fund, the Déné Nation (a Native group in Canada), and Zoocheck Canada, in response to special light goose hunting regulations initiated in Canada.

"The result," Kelley continued, "was that in order to prevent further litigation on the light goose regulations, the U.S. Fish and Wildlife Service withdrew the Conservation Order and announced they would begin preparation of an environmental impact statement on light goose management."

However, the urgency of the situation was not lost on members of the U.S. Congress, and Representative Jim Saxton (R-NJ), chairman of the House Resources Committee's Subcommittee on Fisheries Conservation, Wildlife, and Oceans, introduced legislation in July 1999 that allowed the goose population reduction to continue in twenty-four states while the environmental impact statement was being completed. On November 24 of that year, President Bill Clinton signed the bill.

Thus the Conservation Order proceeded, and during 1998–1999 there was a combined harvest in the Central and Mississippi Flyways of 435,034 birds. The regular hunting season harvest was 637,105, so the total for the twenty-four states was 1,072,139 birds.

The Light Goose Conservation Order will continue to allow the special management action until the environmental impact statement is completed and approved sometime in 2005. It is hoped that the order will accomplish its purpose and the fragile arctic breeding grounds will be saved for posterity.

Humans and Nature: is their relationship a partnership or a battle? We are slowly learning it's a relationship in which Nature will always win. The outcome might not be what we expected—or wanted.

THE LAST ONE?

It's a morning fit for the angels as I climb into Kenny's brand-new deer blind. It sure beats the cedar brush and camouflage cloth I've used all spring and summer, but of course this blind would have to be at ground level. I adjust myself comfortably in the cushioned chair and test its swivel. Tripod could fit right here, and big lens could work at any of the lookouts.

As I glance at the floor, I almost laugh. That rattlesnake damn sure wouldn't have bothered me if he had to swish across a red carpet like this to share my shade. Too bad for him that just one of us could be boss so close together in my blind.

A thermos of hot coffee is on a wall shelf, and beneath it a tiny butane heater. There's even a small "relief" pot with a lid. Then I almost yelp, "Kenny, you've gone too far!"

A cell phone! Right there on the wall. Just waiting to ring as the biggest buck in the universe prances up. I'm tempted to call New York, L.A., and Paris and let the phone bill state my opinion to Kenny. Naw, he's having too much fun with all the new money he made this year. Then, staring at the phone, I realize its truth. Of course, it's here so I can call and get help hoisting and gutting my *two* bucks and *three* does.

The feeder is shooting out corn and milo at regular intervals, and a Western Scrub-Jay zooms in almost on top of a rock squirrel. There's a short contest before the squirrel scurries off with its cheek pouches only half full. I could've filmed it if I had my camera instead of this gun, but Kenny flat laid down the law.

"Forget birds today—you gotta do your part," he insisted. "You were there when the state biologist warned me this ranch is getting more deer than it can possibly take care of, and you know I'm never going to let a deer starve to death on land I own. You're our best shot, so old buddy, your quota is two bucks and three does before the season is over. Period."

Two bucks and three does. I think they would almost let us machine gun 'em this year. There's a deer everywhere I look. Don't know how I missed that one yesterday morning—jumped right in front of my truck.

Trouble is, people are feeding 'em to watch 'em because they're "so pretty and so wild," and now there are so many deer that they're coming into town and eating everything in sight, from rosebushes to petunias. Trouble is, when deer browse all the young plants, nothing's left to grow up and make seeds to replace the old ones. They're eating themselves out of house and home big-time.

That die-off I saw in South Texas—I wish I'd never seen it. Starving is a lingering death, an agony I sure wouldn't want my kids to see. So, yeah, Kenny, guess this is my day off from birds.

"You old fool! What did you just say?" I thought. " 'My day off.' You're retired, remember?" But I couldn't miss the migration, and I couldn't miss nesting season, and I had to record that rare hummingbird, and baby birds are so much fun. My day off. "The truth is, you birdbrain, you're working harder some days filming birds and wildlife than you ever did making a living," I mused.

So it's got me. I'm hooked. Hunting with a camera, showing my videos to the clubs and meetings. I like to hear 'em say I'm a natural, whatever that means. They don't know how many hours I put in on research. Katie says I'm buying more bird books than beer.

Whoa! Here come the doves. Must be about twenty, giving the scrub-jay plenty of room as they light. They've got a stranger with 'em. Looks a lot like the Mourning Dove, but considerably larger and longer. I do know MDs, and this one sure doesn't belong here. Must be an escape. But I think maybe I've seen it before.

A dove flares at it to drive it away, and it responds with a harsh *keck* and then a *kee-kee-kee-kee* as it flutters its wings back in protest. It doesn't give ground to any bird, including the jay, and it feeds in dead earnest much faster than Mournings do, thrusting leaves and dirt aside with a very strong bill.

There's no doubt it's kin to the Mournings, but it's half again as long. The usual dark spot on the neck of the Mourning is missing. Its breast is much redder, and the rump has a decidedly blue cast. Now, that's what I call a pretty bird.

I'm not taking my eyes off of this stranger. My brain is checking every detail, trying to remember where in the universe I saw one like it.

It nods its head differently from the Mourning Doves—more circular. Its legs are red and seem too short. Its shoulders are bigger. *Great balls of fire!* I'll be a whipped hound dog if it isn't just like those Passenger Pigeons we saw mounted in a museum in Florida last year! I made videos of 'em! Showed 'em on my programs. *Damn! Double damn!* Where *is* my camera?

"Stop holding your breath," I reminded myself. "Calm down, calm down. Don't go getting buck fever over a bird—there aren't any of those anymore, remember? They're extinct. Gone. Zero. Zilch. This is some kind of joker, a bird from Asia or some other faraway place that got out of its cage."

But look at it! It's right there in front of my face eating Kenny's milo. It's real, it's *alive*—and it matches every mark of what I got on film.

"All right, birdbrain," I told myself. "Look at the eyes. That's a detail you particularly noted on those stuffed pigeons. Remember, their eyes were red."

My hands are shaking as I focus my b'nocs on the bird's eye. It's *red!*

My heart is knocking like a loose piston rod and I want somebody, *anybody,* here *now!*

If I holler, the bird will fly. If I step out, the bird will fly.

My eye settles on the phone.

"Oh, bless you, my good buddy, Kenny," I'm saying, as I begin dialing, then blurt into the mouthpiece, "I need a witness!"

Well, what would *you* do if you thought you'd found a bird supposed to be extinct?

* * *

Let's shift the scene to Galveston Island on the Texas coast. It's March 22, 1959, and Dudley Deavor and Trevor Ben Feltner are searching for their first Whimbrel (still noted at that time in some field guides as the Hudsonian Curlew).

Both men are seasoned bird observers, and when they see a small curlew alongside four Long-billed Curlews, they assume it is a Whimbrel, but study shows that, instead of being grayish, it is buffy, and its bill is much shorter and thinner than a Whimbrel's. They are excited, dumbfounded, unbelieving, and understandably cautious. There has been no published sight record of the Eskimo Curlew for fourteen years, but they tentatively identify this bird as a species long-thought extinct.

Victor Emanuel, owner of Victor Emanuel Nature Tours of Austin, told me recently, "Ronald Fowler of the Houston Outdoor Nature Club and I were with Deavor and Feltner when I spotted the bird or one like it again on April 5, 1959, in a field several miles from the original sighting.

"It was frustrating. We couldn't get good enough looks at the bird to be absolutely certain it was an Eskimo Curlew. The major problem was that Ludlow Griscom, a world authority, stated in Peterson's eastern field guide that the bird's leg color is a dark green. Our bird appeared to have slaty gray legs, so we were reluctant to pin it down.

"Back in Houston, I simply couldn't get the bird out of my mind, so I began researching and found that both [Robert] Ridgway and [Edward Howe] Forbush stated the bird's legs could be a dull slaty gray or grayish blue. So three days later I headed back to Galveston.

"The weather had been cool and rainy, unfavorable for migration. Nevertheless I was surprised and elated to find the bird still there. Never in my life have I ever studied any bird as hard as I studied the curlews in that field for two solid hours. I managed to flush the bird several times to see the reddish cinnamon axillars and underwing coverts as the bird raised its wings over its back as it alighted."

Some of the excitement of that day came back in Victor's voice as he recalled, "With a thirty-power telescope I even got the curlew in the field of view with a Whimbrel and a Long-billed Curlew—the perfect comparisons!"

Feeling confident, Victor nevertheless needed a confirming witness.

"On April 10," Victor recounted, "Dr. George Williams, an English professor with Rice Institute and an amateur ornithologist, went back to Galveston with me, and as if by miracle, the bird was still in the same field. Old descriptions of the Eskimo Curlew described it as very tame, and as we approached closer in stages with the telescope, the bird showed no nervousness. It fed, walked about, squatted in the grass, and preened itself as we watched. Then a Whimbrel began pecking in the same clump of grass the curlew was pecking. Their differences were so clear and obvious that there could be no doubt we were viewing *Numenius borealis,* the 'lost curlew.' "

For four successive years an Eskimo Curlew appeared on Galveston Island in what became known as the "curlew field." The owner of the field said he had actually hunted the curlews in this same pasture many years past.

Numerous other birders—including George H. Lowery, president of the American Ornithologists' Union; Ernest P. Edwards, author of *Finding Birds in Mexico;* and Edgar Kincaid, editor of the monumental *Bird Life of Texas*—were among the observers who concluded the sightings to be the Eskimo.

Kincaid, always the stickler for accuracy, at first brought up the possibility that the bird could be a wanderer, an Asiatic-Australian species called the Least Curlew, but he agreed it was improbable that such an individual could stray so far from its usual territory for four years in a row. Edgar Kincaid agreed that the identification was "sanitary."

The Eskimo was successfully photographed in 1961 by Charles McIntyre of Houston and in 1962 by Don Bleitz of Los Angeles.

There was no doubt, as of April 3, 1962, that the Eskimo Curlew still lived.

Victor Emanuel (left) *with the revered Miguel Alvarez del Toro of Mexico.*
(Courtesy of Greg Lasley)

In Rockport, Texas, in 1966 I interviewed Judge Allen Simpson of Racine, Wisconsin, and he told me the following story.

"I've been interested in birds for years, so when an Eskimo Curlew was reported in Galveston in 1959, I made three unsuccessful trips there to see it," the judge recalled. "I began studying up on it, including skins and mounted birds, until I felt I could identify it if I ever had a chance.

"Connie Hagar, Mr. and Mrs. John Fiske, Elizabeth Dreher, and I frequently had been taking a lunch to Rattlesnake Point in the Rockport area in the late afternoons to watch the ibis, spoonbills, and egrets fly to roost. Mrs. Dreher pointed across a little pond about a hundred and twenty feet away and asked, 'What bird is that?'

"I was in the backseat with a Balscope, thirty power, and when I put it on the bird, I said right off, 'It's an Eskimo Curlew!'

"Connie had only her binoculars, so she said, 'If it has green legs, it is.' And it had green legs.

"Very excited, we took turns with two scopes for at least ten minutes until another car drove up and the birds flew. Next morning, we went back, and I'm sure I saw the bird again, though I didn't get a perfect look. So far as the records go, that is the last Eskimo Curlew identified alive."

I make a note here that despite these birders' well-earned reputations as skilled and knowledgeable observers, current procedures would have required careful and detailed written descriptions and hopefully a photo before this sighting would have been officially accepted by the Texas Bird Records Committee.

Five months later, September 4, 1963, a hunter shot an unusual bird on the coast of Barbados in the West Indies. Fortunately it was given to Captain Maurice B. Hutt, who put it in his deep freezer. More than a year later it was finally identified in Barbados by James Bond of the Academy of Natural Sciences in Philadelphia as an Eskimo Curlew.

Was this Judge Simpson's same bird, continuing its traditional migration route, and was it the last of its kind?

No, according to a story handed to me that was published September 11, 1974, in the *Wall Street Journal.* It featured Joseph Taylor, a legendary birder, the first to score seven hundred Lifers for the North American List, and the first treasurer of the American Birding Association. Joe and his wife, Helen, had discovered an Eskimo Curlew in Texas (no definite location) in 1966. Joe believed they were the last to see it alive.

"I couldn't even breathe, I was so excited," he recalled. The article furnished no further details.

Joe is now deceased, but I wrote Helen in May 1998, asking her to give details. She replied, "I only kept my notes up to 1961, so I went to Joe's files and couldn't come up with anything helpful." So Joe's record is lost to history unless someone who knew Joe can furnish more details.

There was more good news, however. John E. Lieftinck of Rockport sighted an Eskimo Curlew in the Rockport area on April 30, 1968. Like the other observers, he was able to view Whimbrels and Long-billed Curlews together for comparison of the three species, and as the smaller Eskimo flew, he could clearly see the rich, reddish brown wing linings. This sighting had no other witnesses and no photos, so it had no official recognition.

Birds of North America: Life Histories for the Twenty-first Century, number 347, gives a thorough and detailed history of the Eskimo Curlew. It is surprising to read that from 1945 to 1985 there have been reports of this curlew in twenty-three different years, with a total of about eighty individuals and with the largest a flock of about twenty-three birds in 1981. Some authorities question this last sighting, as it was on a sand spit out in the bay—not at all the type of habitat preferred by this species.

The most frequent sighting reports since 1945 have been from Texas. In 1987 there were four "apparently reliable" sightings made there. The computer records of the U.S. Fish and Wildlife Service contain several unsubstantiated sightings in the early 1990s.

Two unsubstantiated sightings totaling three birds were made in the Prairie Provinces in mid-May 1996. These birds in southwest Manitoba had descriptors matching those of Eskimo Curlew.

The last confirmed sighting of the bird in its South American wintering grounds was in 1939 by Alexander Wetmore. The last specimen was collected in 1963. A systematic and extensive search in the grasslands of Argentina in 1992–1993 failed to find any.

In 1940 Ludlow Griscom had responded to an inquiry from Clarence Cottam of Texas concerning Griscom's report of seeing two Eskimo Curlews in Ipswich, Massachusetts. Griscom ended his reply, "It would be inexcusable to collect an Eskimo Curlew to validate a sight record, and on the other hand, when a bird is practically extinct, is a sight record any good?" Such a collection of the bird would not have been illegal in 1940.

In an interview I recorded with Connie Hagar on February 26, 1966, in Rockport, Texas, we discussed the Eskimo Curlew. Was the Eskimo Curlew of 1952 the rediscovery of the bird?

"No," she said. "Joe Heiser and some others reported one on Galveston Island, and then later I saw one here. When the second one was here, Dorothy Snyder of the Peabody Museum at Salem, Massachusetts, was also here. I asked her if she wanted to see an Eskimo Curlew. Well, it was with the other two curlews, just as shown in [Richard] Pough's guide. We looked at books and studied descriptions when we came home for lunch, then went back, and the bird was still there.

"When Dorothy got back to Massachusetts, the curator there wouldn't let her have it. Said they had been extinct for at least ten years. She wrote back that 'If it's all right with you, I'll take it off my list.'"

Connie continued, "I wasn't familiar with the picture ahead of time, but when it was reported from Galveston, I had begun to study it. Then, with it standing with the other two, it simply knocked you down if you knew curlews. And, of course, the green legs too. I never did get to hear it call. This was the third time I'd had it here. This showed the professionals that it was following the traditional route it used to travel when they used to hunt them. I haven't seen it every year."

Though it nested on the tundra in western Canada and Alaska, when the Curlew migrated in the fall, it flew almost due east to Labrador to feast on crowberries. There it doubled its weight in preparation for its more-than-nine-thousand-mile migration above the western Atlantic Ocean to South America. It was nicknamed "the doughbird" because its fat breast sometimes burst when it hit the ground.

To understand the amazing physical feat of this bird's migration, read *Last of the Curlews,* by Fred Bosworth. Of all the wings of shorebirds, the long, narrow, and gracefully pointed wings of the Eskimo Curlew are best adapted for easy, high-speed flight. Only golden-plovers fly as fast, so Bosworth has his lone (last?) curlew satisfy its flocking instinct by joining a flock of golden-plovers. Larger and stronger, the curlew in the story takes the lead, the most arduous position, most of the time.

I quote Bosworth here: "When the tumbling Labrador hills dropped from sight behind, the last orienting landmark was lost, but the curlew led the flock unerringly on. Somewhere in the cosmic interplay of forces generated by the earth's rotation and magnetic field was a guide to direction to which hidden facets of his brain were delicately tuned. He held direction effortlessly, without conscious effort. An unthinking instinct millenniums old, was performing subconsciously a feat beyond the ken of the highest consciousness in the animal world."

Bosworth, a trained ornithologist, presents an extraordinary understanding of the varying terrain, the ever-changing weather and water as he puts you in the place of the curlew, flying fifty miles per hour, crossing the Gulf of Saint Lawrence in five hours, and passing from the pale green of arctic water to the misty maw of the Atlantic, where five of the plovers are lost to the sea in the cloddy snow, and then on to the indigo blue of the Gulf Stream, the vivid red of plankton, and the Sargasso, the weirdest of seas.

By the time the curlew and the remnants of his flock reach the patches of savanna of the great Orinoco River in South America, they have flown without food or rest for almost sixty hours. They have flown close to three thousand miles on the body fat they stored in Labrador. (It isn't known, but it is possible that the curlew is capable of swimming and may rest on the water as it makes its long migration south over the western Atlantic.)

The drama of the curlew's seeing another member of its own species for the first time, a female, is memorable. The flight of the pair toward the north together is likewise dramatic. I will say no more, except this small book is a must-read for bird lovers.

By including excerpts from scientific reports and journals, Bosworth shows the progression of the curlew toward extinction. Arthur C. Bent, in his *Life Histories of North American Shore Birds,* begins his discussion of the Eskimo Curlew thus: "The story of the Eskimo Curlew is just one more pitiful tale of the slaughter of the innocents. It is a sad fact that countless swarms of this fine bird and the passenger pigeon . . . are gone forever, sacrificed to the insatiable greed of man."

For a population that numbered at least in the hundreds of thousands and was killed literally by the wagonload, salted by the barrelful, canned for use in London, dumped in great heaps in fields to rot, and slaughtered so relentlessly that the birds could not come in to feed, Bent's statement is the chilling truth.

Certainly, this was one valid basis for the antagonism that grew between hunters and conservationists, some of which still lingers even in our more enlightened times when we know there were also other mitigating factors for the bird's demise. Important was the plowing of the midcontinent prairies, which caused the extinction of the Rocky Mountain grasshopper, a food that had been vital to the migrating curlews. The same change of prairie to agriculture took place in South America, and in both places the elimination of wildfires also decreased a favored habitat.

The bird had a low reproduction rate. Worst of all was the bird's sociability. It would respond to the calls of its own injured flock mates and stay near, thus becoming a target itself.

What a gift it would be to have this bird return. As a hunter's wife, I could bemoan its loss as an excellent food source, "delicious eating, being tender, juicy, and finely flavored." Farmers could wish for it as an efficient insect eradicator. As a bird lover, I can only thrill to its aesthetically beautiful flight in the descriptions of those who saw it and can only imagine the music of "the soft, tremulous, far-carrying chatter" of a flock in flight.

Ted Eubanks of Austin tells me that the "Curlew Field" on Galveston Island is still open, as in the past, but not so heavily grazed as before and therefore a less desirable habitat for this bird.

If the miracle of several Eskimo Curlews is reported and positively verified, what could or would authorities do? Attempt to initiate a program of captive propagation for eventual release into the wild? Hardly. Attempt rapid improvement of habitat and protection? Too late? Is there anything that could be done to keep this rarity with us?

I could advise only this: if a verified, genuine, living Eskimo Curlew does show up, no matter where or how, authorities had better have a well-defined plan to manage the adoring birders who will arrive in scores to have for themselves the glory and the thrill of seeing it alive, tame, and perhaps even easily studied and observed and will be hoping to hear its call.

How can I answer with anything but that I would want to be there too?

WISH-BOOK BIRD

You're probably too young to remember the big, fat catalogues from the mail-order stores that were a fixture in nearly every household when I was growing up. We called them "wish books," and they had a long life, being relegated to the outhouse when outdated. I spent hours pouring over them, circling or even scissoring out the items I wished for. I knew I was making believe—but just maybe . . .

In mid-July 1971, when Roxana Rose of Austin and I were cruising the higher points of the rugged Edwards Plateau, we carefully scanned every Turkey Vulture we saw, hoping that our fairy godmother had waved her wand over it, but I knew we were making believe. There have been only sporadic Texas reports of Zone-tailed Hawks, so this had to be a bird only to dream about—a wish-book bird.

Then Roxana, Red, and I parked the Road Roost at Panther Junction in Big Bend National Park on August 6, and in minutes Roxana was shouting my name and "Come here! Hurry!"

"Red is too young to make me a widow," I said to calm myself, so I grabbed my b'nocs, but I didn't need them, for the huge black bird was less than fifty feet high and directly over my head.

A strange thing happened, a sudden empathy between me and the magic flyer. This bird was hanging in the air like a guiding light and spreading his tail in a fan for a perfect display to me of three white bands; then he balanced himself gracefully to lower his black head with its powerful yellow beak and give me a piercing look for what seemed minutes. It was apparent he had come a long distance for the express purpose of viewing me to full satisfaction. (For his Human List?)

When I was again conscious of my feet on the ground, I raced to the service station screaming bloody murder, "Red! Red! Come *quick!*"

When I realized that all of Panther Junction was staring at a female gone berserk, I hollered much louder, "A Zone-tailed Hawk!"

Red managed a departing look as the bird of prey circled away, and then we all lost it as it mingled with some Turkey Vultures and in a whiff became just another common ol' Turkey Vulture with the side-to-side tilting flight of a "TV" and no white bands in the tail. No wonder it is seldom sighted in vulture-plentiful Texas.

Next day, as we neared the Pinnacles on that killing hike up Mount Emory to Boot Springs, even chief naturalist Ro Wauer for a second or so mistakenly picked out a TV as a ZT. But as we approached the hawk's high nesting area, there was no doubt, for the handsome soarer swooped a few feet above us to look all of us over carefully.

This Central American hawk naturally would be less populous here at the northern limits of its range, but this one was here for a specific purpose. It gave me not only a fantastic wish-book premium but also the five hundredth bird on my North American List.

In *Life Histories of North American Birds of Prey,* Bent noted that the white zones in the ZT's tail do not show at all angles and are conspicuous only from below. The bird had given me the best possible view.

I don't doubt for an instant that it came especially to find me, for Ro Wauer said there has never been another record of this species descending the ten miles or so from up in the mountains down to the valley of Panther Junction.

So I'm thinking, everybody should have magical powers at least once in a lifetime.

CROSSED-UP NOMADS ARRIVE

A failing pine crop in the north has brought a winter eruption of inveterate nomads to Texas this November of 1972, and it'll be a lucky streak if you can find them.

Red Crossbills are "unreliable," the personification of here today, gone tomorrow, and this irregularity includes nesting at any season. To make it harder for the Texas birder, they show up here only every decade or so.

The first report in 1972 was of fourteen vagrants in Fort Worth in mid-October, an event noted by newspapers; twenty-four were publicized in Dallas on October 28; and on October 29 at least nine appeared in Ralph Clearman's backyard in Bartlett.

Clearman put them on the Bird Alert, and on the chill, drizzling morning of October 31, Roxana Rose and I found Austinite Jim Tucker already there in a hoping-and-waiting session in Clearman's yard. There is a special suspense in waiting for something that may never arrive, and we allowed ourselves only half hope.

Finally, Jim, certainly one of America's champion birders and the founder of the American Birding Association, moaned, "They're not gonna show," but shortly his alert ear caught a faint feeding call.

The miniature cones on some exotic cedars about twenty-five feet tall were the attraction. Brick-red male, yellowish female, and duller immature plumages were all represented as about sixteen birds enjoyed a treetop brunch less than fifty feet away from us. When they hung upside down and held a cone in their feet, they behaved so much like parrots that I wondered if their feet matched parrot feet, with two toes front and two behind. But, no, they had the same three-and-one finch arrangement.

They pried cones apart with their crooked forceps-style bills and then handily extracted small seeds with their tongues.

Crossbills are not shy, and soon several at a time came down to bathe in a large rain puddle about fifteen feet away from us, often demonstrating their pecking-order dominance of male over female over immature. A soft *kip-kip-kip* conversation kept the members of the restless group in close touch with one another.

"Do all their bills cross the same direction?" I wondered.

"You *would* bring up a question like that!" Jim reproached, but patient study of the lively birds in the telescope proved indeed that some are right-billed and some are left-billed. In case you worry about that sort of thing, bills of baby birds are straight and don't begin to cross until several weeks after the birds leave the nest, so though adults feed the babies by regurgitation, no bills get all crossed up.

Two days later the restless, fast-feeding gypsies were gone from Austin but were reported in Temple, and then, on November 5, Suzanne Winckler and Rose Ann Rowlett of Austin heard and briefly saw a flock of Red Crossbills in Buescher State Park. The colorful, unique birds could linger in the "lost pines" there for a while, since pinecones furnish one of their staple foods.

To catch up with these rovers, don't put salt on their tails. Instead, put salt on a feeding tray, where these sparrow-size finches can eat it. They are so fond of salty substances, they sometimes dine on salt put on frozen highways. They also have been seen flying repeatedly into chimney smoke, apparently for a smoke "bath."

***

I've been asked to define what "Bird Alert" meant back in the 1970s. As mentioned earlier here, my effort to organize an alert using ham radio operators wasn't successful—too many steps, too many people.

There are not many old-timers hanging around now, so it was my good luck that, right here in Austin, Ted Eubanks came up immediately with a method that had been used on the Upper Gulf Coast of Texas. "The telephone tree, of course," he suggested. "Each person had three people on his or her list. Those three had three more on their lists. We could cover the situation pretty well in a reasonable time."

That was early-bird e-mail.

PROBLEMS WITH A TIN EAR

It was during World War I that the family folklore started concerning itself with my singing abilities. I was told that at the age of two I could accurately sing "It's a Long Way to Tipperary" all the way through. Naturally, I took it for granted that I had an ear for music. This myth began to fade when I undertook to sing in the church choir, and it vanished entirely when I began to study birdsongs.

Admittedly, birdsong is a unique and complex form of music. There are hundreds of bird species, each with its own repertoire; species often have local dialects; and individual birds have their own tone and delivery. Yet, even with these difficulties, many people learn and retain birdsongs much more reliably than I do.

I love operas, symphonies, concertos, hot blues, danceable, old-time, and in fact almost any music well done, so I'm not musically illiterate, and with birdsong I manage fairly well in familiar surroundings or if I'm where I have some idea of what to expect. Sometimes in my youth I could even hear the higher frequencies better than my companions, but set me down in strange territory and usually I would do just as well birding by ear on Mars.

"Isn't that a Harris's Sparrow?" I'll ask expectantly, and Red will answer wearily, "That just happened to be a meadowlark. Why don't you stick to fire engines, Hon?"

That man Red has an amazing retention, and he gets downright superior about it at times, so it was good medicine when a plain ol' Bewick's Wren baffled him one summer. Of course, it baffled me too, for never in the years of listening to Bewick's have either of us ever heard this particular lyrical and intricate song.

My conclusion is that a musical ear, alas, is very much like beauty. You'd best be born with it. So I console myself that I know a screech-owl when I hear it—unless it's an ambulance whining in the far distance.

An aid to learning birdsongs is to practice one bird's song and then compare it with another species' song to notice the differences. Using a birdsong recording can give enough of a bird's variations to make such a comparison.

At Roy Creek, Red and I played one of Donald Borror's birdsong records outdoors with considerable volume. Birds came flying in from every direction, including eleven Northern Cardinals. Most unexpected were three Black-chinned Hummingbirds, which lit very close, and two of them even hovered briefly above the record player. Is this the way to call a hummer?

I was in an oak grove one day trying to call in a Bell's Vireo, somewhat rare for our area, when a large shadow passed over me. I looked up to see a Turkey Vulture circling over my head.

"You'll have to wait a while," I told him, and he soared away. I remembered someone said, "Vultures make such nice shade."

I was in awe of Fred Webster, the first person I ever witnessed calling birds. It was at Palmetto State Park. Fred got down in a ditch (never mind the poison ivy) and, well-hidden, began to imitate a screech-owl. He did it to perfection, and in a short while birds were coming from all directions, to the delight of the surrounding birders.

Thus, out in the country near our neighbor's barn, I decided it was time for me to practice, and leaning against a tree, shielded by a bush or two, I did a fair performance.

It was an instant happening!

A western Rufous-sided Towhee (now called the Spotted Towhee) zipped toward me and almost flew out of its feathers, trying to stop before lighting on my head. Equally surprised, I was certain I had jumped out of my own skin for an instant or two. I would grade my performance as a very successful first try.

Through the years, Red and I have used many methods of finding and seeing birds, such as squealing against a fist, using man-made squeakers and tremolos, tapping on bark or posts, shouting, pishing, cussing, and so forth.

It was at the first convention of the American Birding Association in North Dakota that I was privileged to witness the famous ear of the great Roger Tory Peterson at work. This was on a field trip with seven busloads of birders to Longspur Pasture, with the goals of Baird's and sharp-tailed sparrows. If a seven-busload field trip staggers you, it did me also. However, parties of this size were entirely successful in the wide expanses of meadows and potholes because birds were everywhere, there were at least a dozen telescopes, and of course, many of the world's best birders were present to help others.

As royalty, Peterson was given the privilege of coming to the field trip in

Roger Tory Peterson (left) *and Chandler Robbins, the grand masters of birdsong identification. (Courtesy of Bob Danley)*

his own car and of being tardy. Happening to travel in the large motor home that was the emergency first aid station and rest room that trailed all the field trips, I was tardy too.

As I stepped out, Peterson's greeting was, "Do you hear the Baird's Sparrow?"

I didn't. The Master guided me closer and gradually I picked up the three *tik*s and a trill, and I could see the bird, perched on a tall grass stalk, raise its head and deliver its song. Too far to discern the orangey crown stripe—and remember, I'm one of those who wants to see before counting. With bird and song and the Great Ear beside me, I counted it as a Lifer, but felt much better when I had a good look at the orangey crown later.

Later, I recalled a story told me by Ethel M. Gray of San Antonio, Texas. In 1936 Ethel attended one of the early Audubon Nature Camps on Hog Island near Medomak, Maine. Both Peterson and Allan Cruickshank were instructors. Peterson was a strict field trip leader, making everyone walk single file, and no one talked except in the lowest tones or whispers.

The gifted young leader listened in puzzlement, then grew even more puzzled when he saw the singer. Cocking his not yet so famous ear, Peterson shook his head sadly at the bird and passed sentence on the culprit.

Ethel reported what Peterson had decreed: "That bird just isn't singing *right.*"

As time passes and my hearing dulls, my greatest worry now is this: If, as I stand at heaven's gate and I hear a bird singing, will I know it is the lark?

BORN TO SING

It was during a Texas Ornithological Society field trip in Fort Worth that I was first aware of him. As our group was fielding birds with binoculars, a small, elderly man was keeping his distance well away and strangely, without binoculars, facing intently one direction, then with head turned skyward, just as intently facing another.

This was Charles Hartshorne, the philosopher-aesthetician living the ornithological side of his life, doing what he has done in forty countries and at least forty-five of our United States—listening, listening, listening to birds.

Our acquaintance was casual until January 11, 1966, when I called at the Hartshorne home in Austin for an interview. His wife, Dorothy, a stately, gracious woman and the local Audubon chapter president, led me through the house to a study–dining room featuring a window that faced the nearby park. The doctor's book, *Born to Sing,* would not be published until 1973, and I don't recall how I learned of his birdsong studies.

Charles Hartshorne, though small and wiry, emanated a presence as he sat down in an old-fashioned rocking chair. He had taught in eleven universities in the United States, Germany, and Japan. He was named an Ashbel Smith Professor at the University of Texas and soon after was awarded the Lecomte du Nouy Foundation Prize for the "best work contributing to the spiritual life of this epoch."

I was in the early years of educating myself about birds and everything connected with them; I had advanced only enough to fall totally under their spell; I had never had a formal science lesson; my exposure to philosophy consisted of reading and rereading a book by Will Durant. Not unexpectedly, I trembled as I sat down to interview the Great Man.

Soon I realized I was talking to the doctor's third persona, Hartshorne the inventor.

"I accepted Eliot Howard's territorial view as the main function of song," the professor said, taking charge, "but I was trying to find out what other functions or factors would be involved in some birds singing much better than others."

His extensive study of birds had led quite naturally, it seemed, to an aesthetic and scientific system for evaluating their songs, which I was relieved to learn was in a logical form that I could not only understand but also appreciate.

"Did you do your studies in connection with your teaching work or on your own?" I asked.

"I am the only person I know of who received three Fulbright scholarships," he replied. "I was interested in and I've written about *all* the creatures that sing—insects, frogs, whales, and so forth, but birds are the most conspicuous. The grants didn't cover it all, but I was determined to do this work, so I used my own money when it was necessary. My wife, Dorothy, went with me most of the time.

"The so-called primitive birds don't sing; it is only the highly developed birds that do," he continued. "I was just considering that singing is an evolutionary development and achievement, and achievement has degrees. Miles North, an Englishman in Kenya, gave me some hints, including the use of numbers, but he didn't develop it further.

"What I did was to find definite criteria by which singers can be objectively graded and thus compared on a scientific basis. The question was how far the idea of degrees of ability could be taken.

"There would have to be more than one dimension in comparing musicians, and one criterion would be *loudness*. This must be qualified by the habitat from which the bird sings and by the relative size of the bird, for a small bird can't create the volume of a large bird. Thus judging is done on how far the sound carries and how effective it is at a distance. Sometimes a bird, such as the California Jay [Western Scrub-Jay], will sing a little song, a 'whisper' song, so soft it can be heard no more than twenty feet away. Such a song can't have much biological importance; he probably is just amusing himself."

"I heard a female cardinal performing a beautiful song like that," I ventured, proud to join the conversation.

"Is that right?" he replied. "I have never heard the female sing. Was it possibly a young male?"

How could I answer such a question? I sat like a frightened stone, ashamed of being nakedly uneducated.

"It's true," he went on. "In this species both sexes sing."

141

I relaxed in my chair.

"Another measurable dimension is *continuity:* the ratio between the amounts of time the bird sings and is silent. Since a bird's memory span is very short, a minute is a sufficient interval for grading this. One comparison is a Henslow's Sparrow, which sings a song lasting only a fraction of a second and with pauses in between that can be fifteen times as long as the song. In contrast, a Sky Lark in a one-minute interval will sing continuously."

"I heard a Whip-poor-will sing constantly, just on and on," I recalled.

"Yes." He paused for a moment. "But that bird, of course, is not an oscine. It doesn't have a highly developed organ for song."

I realized as I sat here in this chair that at this time silence was golden for me.

"Another measurable dimension is *scope,*" Hartshorne continued. "This includes variety, complexity, and versatility—every kind of contrast. I apply this always to the individual of one species as compared with the individual of another species, for careful observations show that even within a single species any two birds will sing at least a little differently. One example of scope would be pitch contrast: for instance, singing a note two octaves higher at one time than at another. Under scope, a bird with one song of simple pattern would be compared unfavorably with a bird that has as many as forty little song patterns, a repertoire."

I was much impressed. "You must have a strong musical background to have conceived all this," I marveled.

"No. I am deeply responsive to music, and I have heard any amount of good music, but I have no musical training. It is one of my greatest regrets. I took a course a couple of summers ago, trying to help myself, but it's one of my worst weaknesses.

"Let's continue," the professor said. "One of the more subjective dimensions of birdsong is *imitative ability.* The [Northern] Mockingbird and the lyrebird rate very highly on this. The mocker can imitate as many as forty species. I think that all songbirds have some capacity to imitate, and I give them all at least one point in this category. Even a House Sparrow, which is a poor singer, has learned to imitate other birds when it has been kept in captivity and has nothing to do but sing.

"However, a bird can imitate only other birds it hears, so a judge must necessarily be familiar with the birds of that region. Some imitations are not very good and thus are subject to interpretation by the listener. Bedichek was wrong, but very intelligently wrong, about the mockingbird not imitating. I am sure he would have taken a different position if he had had a wider geographical experience. No one can expect a Texas bird to imitate a bird that is not in Texas—for instance, a towhee."

"Oh, we have towhees in Texas. Lots of 'em." *Aren't you getting too brave?* I asked myself.

"You do have towhees? Where?"

"My very first one was in a brush pile in Pease Park in central Austin. It was so beautiful, I scarcely believed it was real. And it was calling." *Aren't you growing smart too fast?* I thought. *Watch out.*

After a pause accompanied by a quizzical look right through me, Hartshorne commented, "Well, let's put it this way. I have never heard one in this area.

"*Tone* is another criterion for judging birdsong," he said. "There is a definite physical difference between a noise and a tone. A noise is a mixture of pitches, whereas a tone is a single pitch, plus some overtones. This can be measured instrumentally. Nearly all musical birds have some unmusical alarm or scolding notes. The exquisite pure tones of the Wood Thrush contrast forcefully with its scolding chatter. The guttural croak of a Hermit Thrush is noisy. The nightingale has a very harsh scold, certainly a noise."

"Would you call those sounds warnings?" I asked.

"Yes, or even weapons," he agreed. "That's a general understanding. Birds often have what could be called a song duel."

Dorothy brought us each a cup of tea. Now Hartshorne the scientist resumed, "The most subjective and difficult category of song evaluation I call *organization,* getting coherence in a musical sense—the musical design. Bird music is very simple, but every simple musical device is used by birds, so it is an objective fact that birdsong is musical; however, there is no evidence that a bird can keep track for more than a few seconds of what is happening musically. A human can keep track of a pattern of a half hour.

"Even so, there is proof that the bird is aware of musical organization because some sing the same song in different keys or even at two different speeds. Very few birds are capable of the latter feat, but quite a few are able to sing a trill both fast and slow. Other devices they use that show musical organization are harmonic intervals, fifths, octaves, accelerando, crescendo, diminuendo, and so forth."

"Did you do your own recording of all these songs?"

"I did some, but I often used recordings from others," he explained. "One thing that is somewhat puzzling is that when songs are played at slow speed, a great deal of detail is disclosed, and there is fairly good reason to believe the bird is aware of this detail. So although the songs are simple in one way,

On a Texas Ornithological Society field trip. Left to right: Edgar Kincaid Jr.; Charles and Dorothy Hartshorne; and Victor Emanuel. (Photo by Suzanne Winckler)

measured in time, in terms of the number of notes and the number of musical intervals they can be fairly complex. But there is a definite limit—nothing like a symphony going on."

In all six categories, Hartshorne gave a score ranging from 1 to 9, so that a bird's final score looks like a six-digit telephone number.

"As to personal taste, I noticed you didn't rate any song in terms of how much you like it," I commented.

"Yes, a song can be rated high, yet one may not like it personally."

I had one final question for him: "How did you get interested in birds?"

"I was sent off to boarding school in Pennsylvania when I was about fifteen," Hartshorne recounted. "It was in true, elegant country with sheep farms, woodlots, and woods. I discovered Reed's guide to song- and insectivorous birds, and I also found about a five-power field glass for five dollars. I was equipped. The book was devoted largely to songbirds, which was very fortunate, for it made me specialize right in the beginning."

The interview was over.

I did it! I didn't disgrace myself. Now, if I could only get every fact straight.

***

Fast-forward to February 27, 1997. Now the house is almost hidden by trees. Dorothy died in 1995. Charles Richey, a scholar and the doctor's live-in assistant, told me to bring the car into the driveway to avoid a $50 parking fine (high parking fines were a result of the explosive growth of the University of Texas).

Using his walker, the professor himself slowly led me through the house to the big room. It hadn't changed much, except for the larger amount of books, papers, and memorabilia. The same rocker was near the window, where the view was now obscured by invading bamboo, but he chose for us to sit at the cluttered dining table.

Charles Hartshorne, a wisp of a man wearing a Gandhi cap similar to the one Nehru always wore, sat erect. A yellow pad lay open, revealing notes in numbered paragraphs.

"You're working?" I asked.

"Not so much anymore." His voice is thin and quavering. "Writing soon tires me out."

Memories were coming back to me. "Do you recall the dinner you and Dorothy gave for Sewall Pettingill? Red and I were here."

"No, I don't remember that. Sewell was my only ornithology teacher," he noted.

"It has been more than thirty years, Charles, since I first interviewed you about birdsong, and twenty-five since *Born to Sing* was published. Have your theories and premises been generally accepted?"

"Accepted by some, ignored by many. No one has tried to refute them. It is the only book that discusses all the types of creatures that sing. It has never gotten a bad review. I am satisfied with it. Behaviorism is more qualified now. My views have gradually become noticed. I do not complain."

"Would you use the same criteria today?"

"Yes," he responded.

"What do you think of the game and sport of birding?"

As he had grown older, Charles had lost much of his hearing and had to give up his song studies. Ornithology's loss had been philosophy's gain.

"Birding?" He was impatient. "I was working as a scientist. Trying to add to knowledge and understanding, not someone trying to beat a champion with a big list."

"You were very close to Edgar Kincaid," I noted. "In fact, he died in this house. What do you think is his legacy?"

"Edgar was one our greatest scientists," Hartshorne said. "He had a remarkable memory and a sense of organization. Editing Oberholser's three million words into two volumes was a remarkable feat. But Edgar got spoiled."

"What do you mean?"

"He found himself required to take on the financial responsibilities of his family's estate," Hartshorne explained. "It meant he could no longer do what he wanted to, which was to devote himself to birds."

Charles and I were friends now. We were members of the same church, and it had pleased me that through the years he often chose to sit by me to talk about birds.

Several years beforehand, his life was in danger. I remember the bright look on his face when I walked into his hospital room.

"I knew you would come," he said. Richey was there with him, and together they were working on the newest book.

"It's the book that's keeping me alive," he said.

At the time of my 1997 interview, Charles Hartshorne was called the greatest living metaphysician, and in the encyclopedias his name is placed alongside the likes of Dewey and Einstein. On a certain level Charles and I shared some of the same beliefs, but we met best on the subject of birds.

Our church was honored to have Charles Hartshorne amongst us, and the theme of our regional conference that year was process theology, that part of philosophy where Charles had made his greatest contributions. The conference was his hundredth birthday celebration. Another celebration was to bring admiring scholars from around the country and even from Canada.

For the church conference, Red and I were chosen to give recognition to the ornithological part of Hartshorne's life, and this was the reason I was interviewing him again.

"Life is an aesthetic problem first, last, and always," he told me. "It is a moral problem only part of the time. How one views the world depends a good deal on how one views the parts of it one knows best. I know people and birds best.

"We can learn wisdom from birds. In some ways they are more like us than the other mammals. For example, they are bipeds. Many are monogamous, with some deviation on both sides. Birds are born to sing. We are born to talk. I should have said this long ago but failed. I should have written a book, *Born to Speak.*

"Living is deciding, and each creature must do its own deciding, its own living. We are interested in birds because of the dedicated way they show their interest in their own affairs.

"We humans are most catholic in our ability to enjoy other forms of life. Deity is the eminent level of this ability."

A few months later, I was back visiting him again. I'd learned that it was most convenient to go to the side door. I knocked and knocked, and no one answered. Was he all right? I went inside, calling, "Charles? Charles?" No answer.

I hastily explored and there he was at the kitchen table, eating a banana. I patted his shoulder in relief. "I've brought your book," I said. "Will you autograph it?"

A surprised and pleased look came over his face, but he was in no hurry as he finished his banana. Slowly we went back to the big room, and he sat down with the book in front of him.

"I've almost forgotten how to do it," he said, but it was open at the title page. The tremor was so great that no handwriting expert could ever copy the signature, which starts "To Marjorie Adams" and ends with the date.

"You are a precious darling," I said, and my arm was around him. How often do we have the chance to hug a hundred-year-old bird lover?

Charles Hartshorne died October 9, 2000, at the age of 103. He was a proponent of an activist God.

A PERILOUS JOB?

He walked jauntily onstage, and with what seemed sleight of hand he suddenly was holding the outstretched wings of a huge bird. Black and wide, each wing was scored with a long, white triangle.

There was a sighing "Oh-h-h-h-h" from an unbelieving audience.

"How would you like to see *this* sail overhead?" John Borneman asked. "And is nine feet big enough for you? Well, you can see it if you go to the right place and are lucky."

This was John Borneman's introduction to his audience at the National Audubon Society's annual convention in Sacramento, California, in 1967. His replica of the California Condor, realistically created with layers of felt, was the introduction to us in the audience to one of America's rarest species, the California Condor.

"I had read in some of the newspapers that the Audubon Society loved birds and hated people," John deadpanned, "so when I first took this job as condor warden, I didn't put the Audubon patch on my uniform. Several kind people had reminded me it would make a good target.

"I didn't know what to expect on my first call out on the job. I bumped along a backcountry road up to a ranch house. A husky young man came out, glowered at me like I was the tax man, and asked, 'Whatta you want?'

"I took a deep breath and, with my most careful best manners, explained my job.

"The young man looked me over with little admiration, then turned and hollered, 'Hey, Pa, the buzzard inspector is here.' "

In spite of his jokes about it, Borneman's job, like other law enforcement work, could be dangerous. Fierce local resentment centered on the idea that protection of the last sixty or so condors in North America would rob people of hunting and fishing privileges or hinder commercial and petroleum development in the area. In fact, some years before, a plot to kill every one of

(Courtesy of Hopper Mountain National Wildlife Refuge)

those "dirty buzzards" had been uncovered. It was rumored that the plot even included a plan to pay the killers' fines.

Yet cheerful, outgoing John Borneman was winning public opinion. He spoke only on invitation, and so far his audiences had included more than thirty thousand schoolchildren and certainly that many adults. Even so, ignorant shooting of the giant birds continued to be a great problem.

What was most threatening to the condor protection project at that time was a proposal to construct two dams on the Sespe River, the only river left in Southern California that was still wild and undammed. The project would include recreational activities that would attract large numbers of people, which in turn could provoke possible conflicts with the condors.

Humor was one of John's tools. A week or so later, when Red and I met John at a service station, we heard him saying to the station attendant, "Yeah, we're having a bunch of pretty little red stocking caps especially knitted for the condors. We're gonna do our best to keep their bald heads warm."

"The truth is," John told us, "condors can pull up the feather ruff growing around their necks to help cover their heads in extreme cold.

"A lot of people think condors are ugly when they see them perched," he continued, "and they do seem clumsy and ungainly when they walk along the edge of a cliff, and then they'll extend those enormous wings out, drooping

149

at the tips, and launch into space, and instantly they are transformed into a creature of total grace."

As we bumped along in four-wheel drive up the public access corridor of the Sespe Condor Sanctuary in Ventura County, we learned that John's deep dedication to his job was no accident. He felt "called" to the work by his religion and quit his job as a singer with a successful national dance band to give his wholehearted time to conservation.

The long drive gave time for more stories, and John grinned as he continued: "I started the one that we're going to strap little battery-operated heating pads on the branches of the condor roosting trees. We don't want their big feet to get cold either."

He laughed. "Probably the worst story was about Fred Silsbee with the U.S. Fish and Wildlife Service. He told his little girl they were eating a condor for Thanksgiving dinner, and she ended up telling this to the neighbors. What else can you expect of us condor cuddlers?"

At the end of the so-called road, we began hiking a good distance up to a high point, where we overlooked a long ridge running away from us. Very soon John was urging us, "In the distance. There they are!"

The giant birds were rising over Whiteacre Peak, the first two at quite a distance. Then another two came over, about a quarter of a mile away.

This is considered a good view, and it was. Weighing twenty pounds or more and measuring more than four feet from bill to tail, they were a significant bird package. Add almost ten feet of wide wingspan, and they are indeed a spectacle for the ages.

And the ages are what they represented. How often do we stand on a height to look across a great valley and see a living species that existed during the Pleistocene epoch, a bird that ten thousand to twelve thousand years ago could have regularly feasted on the carcasses of giant bison, primitive horses, mammoths, mastodons, giant sloths, giant beavers, ancient camels, and other remains of prey killed by the saber-toothed tiger or the dire wolf? We were looking at a bird of the ice ages.

As we gawked, the words came—so solid, so wide, so flat, so steady, so *huge*. These birds can achieve a flight speed of about fifty miles an hour and can cover as much as 150 miles in a day, but watching them was like watching one of the old, slow bombers of World War II. For the entire time we surveyed them, not one bird of the four flapped its wings. Not once. They didn't have to. Moving their powerful wide-spreading primaries in delicate synchronization with the air, the birds were constantly changing course with minimum effort. Certainly no bomber could be as glorious as these birds, with their full stretch displaying their striking black-and-white pattern.

We were too far away to hear the swishing sibilance their primaries made as they played the wind. It is said that when perching condors shake their feathers before taking flight, they make a small thunder.

John told us more: "They can be really curious about people. Their eyes are red and farsighted, but they may want a closer look and come in right up above you, all the time turning their orange heads from side to side to give you a thorough inspection. Of course, that's also when they make the easiest target.

"But they are entirely different in their nesting area—extremely sensitive to human disturbance. They might not go near their young again for forty-eight hours afterward."

Red and I could see no sign of water in the sweeping valley below us. "We've seen Turkey Vultures bathe," I recalled. "Do the condors bathe too?"

"Yes, they certainly do, and they seem to thoroughly enjoy it. As scavengers, they often have to reach inside a carcass, and that's the reason they have no feathers on their necks. After a meal, they will wipe their heads and necks on the grass and then get a good bath. One of the sanctuary areas the condors favor the most includes a waterfall."

The last four of the rare condors that we saw with John Borneman were over the Agua Blanca Creek drainage. We had a total of eight sightings, for a probable total of four birds. That count was an estimate, for at that time no birds wore tags.

January 21, 1967, was indeed a day to be remembered.

A few weeks later, I interviewed Kenneth Stager of the Los Angeles County Museum. It was his studies that proved that Turkey Vultures have a sense of smell, and he passed along the following insight of the condor, a bird he was very familiar with: "When the birds have gorged themselves, they are so heavily laden with food that they have to trundle along perhaps a considerable distance from the carcass and walk up a slope. Then they'll face into the wind, and, like a giant bomber, they rev up their engines and come running down the slope to gain enough airspeed to launch into the wind and become airborne."

How could we know that almost twenty years later there would be only twenty-two California Condors left in the world? And how could we guess that on Easter Sunday, April 19, 1987, the last condor living in the wild would be taken into captivity?

On that day AC-9 (Adult Condor 9) flew down to feed on the carcass of a fetal dairy calf planted there by biologists with the U.S. Fish and Wildlife Service. Trap nets were released, and the biologist took the big bird into captivity. AC-9's mate, AC-8, the last female California Condor in the wild, had

Chumash rock art of the California Condor. (Courtesy of Keith Allen)

already been captured on June 5, 1986. The future of this magnificent species was now entirely in the hands of humans.

There is an old legend that "as long as condors range the skies, the gods will find their way back," and alongside the biologist to help in this historic capture was a spiritual leader of the Chumash Indians, who call the condor the Spirit Bird and consider it sacred. The Chumash have continued their involvement in the condor program, and a member of the tribe is present nowadays to pray for a condor when it is released to the wild.

The huge bird's ancestors ranged ten thousand years ago over coastal regions of North America, from British Columbia to Baja California, east to

Florida, and now fossils dated at eleven thousand years old show that the range extended even to New York. But by 1900 the condors were restricted to Southern California, and here they were threatened by shooting and poisoning, high-powered wires, and even the collecting of their four-and-a-half-inch pale blue or green eggs.

A condor sanctuary of twelve thousand acres was set up in 1937, another thirty-five thousand acres was set aside in the Sespe Wilderness Area of Los Padres National Forest in 1947, and it was expanded to fifty-three thousand acres in 1951.

A woman named Belle Benchley conceived the idea of breeding captive condors, and in 1952 the San Diego Zoo was granted permission to take California Condors and eggs for the establishment of a captive breeding flock. The hope then was that captive-bred condors could be released to join the wild population and learn how to live wild from the older birds.

However, this plan ended when the permits were revoked after a protest from the National Audubon Society. And it might be said that right here is where the great "condor war" began. In 1953 the first legal protection of the condor was passed in California, stating: "It is unlawful to take any condor at any time in any manner."

This war has been one of the longest and most hotly contested wildlife controversies in America. Combatants included hunters, ranchers, landowners, the U.S. Forest Service, the National Audubon Society, the California Department of Fish and Game, the U.S. Fish and Wildlife Service, the San Diego Wild Animal Park, the U.S. Congress, the Chumash Indians, the Los Angeles Zoo, the Bureau of Land Management, and even some who thought the condor should be "allowed to go extinct in dignity."

When it reached the point that citizen protesters chained themselves to the front gate of the Los Angeles Zoo, the situation was so dangerous that for three months Mike Wallace, a biologist in charge of the zoo's condor program, slept by the "condominiums," where the captive condors were housed. Meanwhile, as humans battled, a majestic antiquity handed down to us through the eons was disappearing from the wild world, the victim of bad politics.

The war ended when it was verified that hunters' or vandals' bullets were not only killing condors in flight but also killing condors that fed on unretrieved deer. Lead poisoning and baits set out to kill animals such as coyotes were a major cause of the condors' demise.

In the wild, condors had been laying one egg every other year. The good news was that, in confinement, they were laying as many as three eggs in a single year. In 1988 the first condor baby struggled out of its egg to be born in captivity. Named Molloko, a Native word meaning "condor," the baby bird had taken two months to hatch. It would never see a human and would be fed and handled by a puppet made to look like a condor.

In 1991 the Peregrine Fund's World Center for Birds of Prey in Boise, Idaho, joined the Los Angeles Zoo and the San Diego Wild Animal Park in the breeding program, and as of December 2004 a total of 132 chicks had survived. Some of the adult birds were allowed to raise their own young. Detailed computer records have helped make appropriate matchups, and the once-feared effects of inbreeding have not appeared.

The aim has been to get as many condors out in the wild as quickly as possible, but there was difference of opinion here also. Two or three times as many birds can be raised with condor puppets as can be raised by parent condors, and it was argued that it would be better to get higher numbers of the birds and then let the environment test them.

Two California chicks were released on the Sespe Condor Sanctuary on October 10, 1991. Along with them were two Andean Condor chicks to act as surrogates, but the latter soon were recaptured when the native condors seemed to be doing well.

However, one young condor managed to drink some discarded ethylene glycol and died. Six more young birds were released to join the surviving chick, but three died after collisions with power lines. The death of so many required the recapture of all the remaining birds.

Obviously, a program of "aversive training " was needed, and electrically charged power poles were installed in their cages to give condors a minor shock when the birds touched them.

All the chicks had been raised so that none would be imprinted on humans. Now the first humans they had ever seen were rushing at them, screaming and shouting, throwing nets over them, and shoving them into cages to be locked up overnight. Birds released after this training did avoid power poles and did seem to avoid people.

The Ventana Wilderness Society and the U.S. Fish and Wildlife Service joined the effort and have released a total of fifty-one condors in the wilderness at Big Sur.

The glad news is that in 2002, for the first time in eighteen years, condors began raising young in the wild. The three sets of parents were all captive-raised. Unfortunately, none of their young lived to maturity.

Lead poisoning from the shot used by hunters has been a big problem not only for the condor but also for waterfowl and other wildlife, so lead shot has now been prohibited for hunting waterfowl and also on certain lands on a case-by-case basis. However, lead-containing ammunition is still used to kill deer, so the U.S. Forest Service has begun a program to increase public aware-ness of this problem, and it is hoped that hunters will make greater efforts not only to retrieve their trophies but also to bury gut piles (the remains of field-dressed game).

The consensus has been that if California turned out to be too dangerous for the California Condor, other localities would be found, and September 23, 1993, seven condors were flown on a U.S. Air Force C-141 Starlifter to the World Center for Birds of Prey in Boise, Idaho, to establish the world's third breeding facility. Five other condors were flown to the same destination in a specially equipped Boeing 727 provided by Federal Express. The reason for using two planes was that these twelve birds were 15 percent of all the condors in North America.

On February 8, 1995, six aversion-trained condors were released at Lion Canyon in Los Padres National Forest, California. The birds used only natu-ral perches and showed improved wariness toward people. In May these six independently found a sheep carcass, their first food not supplied by humans. An additional eight condors were released in the area.

The same year, nine "conditioned" birds were released at Vermilion Cliffs, Arizona, and six more were released in 1996. One of the condors was killed by a Golden Eagle, and another presumably by a coyote. In 1999 one was shot in Grand Canyon National Park. The criminal who shot the bird was found guilty of violating the Endangered Species Act and was sentenced with a penalty fine of $3,200, forfeiture of the gun, one year of supervised proba-tion, and two hundred hours of community service. At Lion Canyon another condor was shot in the foot and died later of complications.

The U.S. National Park Service is also participating in the condor resto-ration program, initially releasing six juveniles in Pinnacles National Monu-ment in California.

Today's captive-bred and released condors, like their remote forebears, seem attracted to humans, and they are apt to invite food from construction crews or to invade campsites. They are big and powerful, and though it is hard to believe, they did break through a screen door and clutter up a private residence.

It was decided that Condor 125 was the ringleader in this and other be-havior destructive to property, so he was captured and held for a year. Since his rerelease he has behaved and has become well integrated with his group at Hopper Mountain National Wildlife Refuge in California.

The last sighting of a California Condor in Mexico was in 1930. Now the condor restoration program has been extended to Mexico's Baja California, a historical condor range. In 2002 six condors were the first to be released in a remote area of the Sierra de San Pedro Mártir National Park. Among them was Xewe, an eleven-year-old that had lived in the wild and could teach the youngsters how to live in the wild and find carrion. To start, they were given cow carcasses on a biweekly basis, but eventually they gained proficiency in finding bighorn sheep, deer, and cattle carcasses. Birds that have been released at this site could fly along the mountains to cohabit with American birds.

As of December 2004, there were 244 condors in the world, all of them vaccinated against West Nile virus and 99 of them living in the wild: 47 in California, 47 in Arizona, and 5 in Mexico. The goal is a self-sustaining wild population of at least 150 birds in California and 150 in Arizona by 2020. The public can watch for condors from a number of observation points. A fire lookout in Los Padres National Forest has been restored and is now a visitor center, with a campground nearby. For other locations, inquire at the U.S. Fish and Wildlife Service or the organizations named here.

Over these many years about $40 million has been spent on condor recovery efforts, including captive breeding, habitat acquisitions, and surveillance. Much that has been learned can be usefully applied to other disciplines. "But it has always been controversial," Ted Molter, spokesman for the Zoological Society of San Diego says. "However, there is no longer an environmental group as there was in the past that advocates letting the species go extinct with dignity. And being able to put these animals back into the wild is a huge step forward scientifically." For many people, the proven power of saving a species and putting it back in the wild not only borders on a miracle but also is worth all the effort and every penny.

There's more to my condor story. Our son, Lew, in about 1976 was flying his Piper Comanche aircraft north of Frazier Park, California, en route to San Luis Obispo. At a distance he saw what he thought was another aircraft directly ahead. As he flew closer, the object appeared to be a small Cessna cruising at the same heading and altitude. Since the ceiling for such a small plane is normally only 6,000 to 8,000 feet, Lew was curious as to how the pilot got it up to a cruising altitude of 10,500 feet. With the two aircraft converging rapidly, Lew expected the pilot to divert his course, but then he realized that the "plane" was actually a California Condor, flying at an altitude of about 10,600 feet. Hardly believing his eyes, Lew slowed his aircraft and proceeded to fly a complete circle around the huge bird. The condor was not alarmed and stretched its neck to look at him as it comfortably cruised westward in the warm thermal air.

The big condor news story of 2000 occurred on April 4, when AC-8 (the last female condor remaining in the wild when she was captured back in 1986) was also the first condor to be released back to the wild. During her nearly fourteen years in captivity she became the mother of twelve offspring born in captivity and the grandmother of many of the young condors released to the wild—a "founder bird" I would certainly designate a hero.

AC-8 had been living for several months at the Los Angeles Zoo in a condominium flight cage one hundred feet long and sixty feet high, with two young male condors born in captivity. The hope was the trio would not only strengthen their flight muscles but also grow accustomed to being together and would remain together, with AC-8 acting as a mentor, when the trio were released deep inside the Sespe Condor Sanctuary.

"AC-8's release was a pretty emotional event for those of us who had worked with her for so long," Mark Hall, wildlife biologist with the U.S. Fish and Wildlife Service, told me. "Sun Air Aviation donated her helicopter flight to the sanctuary area. She was very agitated from this flight, and although the holding pen on top of a cliff is made to look like a cave, she kept trying very hard to escape. However, she finally calmed down and didn't hesitate to let the young condors know who was the dominant bird.

"From this cave the three could look out at the surrounding countryside below to familiarize with it for about a week. This area is where she used to nest and raise young, and when we opened the cage it was only seconds before AC-8 took off, flew around the cliffs briefly, then lit in a dead pine tree. Soon the young ones followed her. All of the birds needed to strengthen their flight muscles, and AC-8 stayed in the vicinity, but on the third day off she went.

"Her return to the wild was amazing. She visited all her old territories, and, most remarkable, she found all her food on her own. On May 2, Mike Barth, supervisory wildlife biologist with the U.S. Fish and Wildlife Service, got a view of her near Glennville and reported she had a full crop and had roosted with a large number of Turkey Vultures in the area."

For many people, it was an emotional event in 2002 when Adult Condor 9 was released back to the wild. He had been the last condor in the world still living in the wild when he was captured in 1987. Witness to the event was a crowd of about 150 observers, and among them was John Borneman, who had guided Red and me to see our first condors in 1967. A television crew and newspaper reporters covered the scene as AC-9 quickly reestablished himself in the hierarchy of condor society, and he explored farther north than any other condor since the 1980s.

As time passed and more condors were released, they all seemed to retain an interest in the release pen where they had been freed, and when AC-9 was still in the pen before his release, AC-8 came and apparently recognized him as her former mate in the wild. She hung around quite a bit, and when he was released May 1, 2002, she joined him. However, he didn't seem to respond, so in a few days she flew off to be with other condors.

Today AC-8 is spending most of her time in her old roosting area. All released condors are outfitted with colored and numbered tags on their wings that don't interfere with flight. They are also equipped with special transmitters donated by Microwave Telemetry.

Thus the California Condor again lives in the wild, and individuals of the species can live forty years. Under the rules of the American Birding Association, no one could have known how many years a birder might have to wait to see a condor that was wild, free, and self-sustaining without the aid of humans and therefore could be counted as a Lifer. Now with Adult Condor 8 and AC-9 apparently successfully returning to be the independent wild birds they once were, the big question for birders has become "Can I count these formerly wild birds for my Life List?"

I put this question to Father Tom Pincelli of Harlingen, Texas, chair of the ABA Recording Standards and Ethics Committee. His answer: "All California Condors wear radio transmitters, and this includes formerly wild Adult Condors 8 and 9. At this time, they and all other condors remain uncountable for an ABA Life List, as they are all under the supervision and aid of several government agencies. However, the situation will come up for review in the future."

Of course, a die-hard birder wants to argue, "It's true that AC-8 and AC-9 wear radio transmitters and have big tags on their wings, *but* they were countable, wild birds until 1986. They have returned to their old territories, have taken up their wild ways, and could possibly even mate again and bear young. Why can't they be counted as long as I see their AC-8 or AC-9 tag?"

I am reminded here of an incident that occurred in 1986. A gentleman named Alcott Conbar Aate, while on a business trip to Colombia, ventured into the countryside and was kidnapped by one of the several rebel factions operating in that country. A demand was made for ransom, and Aate, being an American citizen with all the rights attached to that status, had U.S. government agencies working to achieve his release.

However, after a short time, there was no further word from the kidnappers. Indeed, Aate and anything concerning him faded into the wilderness. Years passed, and Aate was declared legally dead.

Then, remarkably, in 2001 a man claiming to be Alcott Conbar Aate appeared in his hometown in the United States. DNA tests proved he was indeed that person. Despite his long absence, and although he had even been declared legally dead and would have to rebuild his life, he was still an American citizen with all the same rights, privileges, and responsibilities he had held in 1986. Yes, he could even vote! And, yes, I did make this story up.

Perhaps the ABA Recording Standards and Ethics Committee should look hard at three words in connection with these two birds. With their capture, the species was extirpated *in the wild*. The wild birds themselves were not extirpated, and, though held captive, they were still untamed, wild birds and their radio transmitters record that they have successfully returned to their wild life.

But don't hold your breath for a decision from the ABA.

The good news is that as of January 2005 there are a total of 245 California Condors in the world. The captive breeding program has had a phenomenal 90 percent success rate. New generations have reached breeding age and are producing young. Even AC-9 is a new father.

For AC-8 there is a different story. With permission from the landowner and with the help of hunting guide Cody Planck, Mark Hall followed the radio signals to find Adult Condor 8 dead on February 23, 2003, a victim of gunshot. Her body was still resting in the tree where she had died. Planck climbed the tree and carefully retrieved her body, which had sustained little damage. The Chumash Indians have asked for it. No decision has been made.

Her twenty-nine-year-old killer, a resident of Tehachapi, California, was charged and sentenced to a fine of $20,000 and five years' probation. Besides forfeiting the gun and scope used to shoot her, he was barred for five years from hunting and from being in the presence of anyone engaged in hunting, and he was required to perform two hundred hours or community service.

As far back as the 1840s, we have been told that the California Condor was headed for extinction. At the hand of humans, extinction almost occurred, but also at the hand of humans the species still lives. Even more surprising, humankind's modern technology has revealed that this bird is related to the stork family.

I like to dream fancy, expansive dreams, and I'm wondering if, at the passing of the traditional seven generations, one of my great-greats might look up in the sky (crystal bright blue, no smog) and see a giant California Condor, *born a native Texan,* soaring up there on her monstrous wings and looking down with that ruby red eye to reclaim the land where her ancestors used to live centuries ago.

That's a dream worth dreaming.

ANOTHER FIRST FOR TEXAS

What is it like to discover a bird species that has never been recorded on the North American continent north of Mexico? Following, in her own words, is a report from Nila (Mrs. Ben) Copeland of Austin.

"August 25, 1969," Mrs. Copeland recalled, "just another Monday, I think, as I go through the routine of filling birdbaths, replenishing sunflower seeds (spilling plenty for rock squirrels), and refilling hummingbird feeders.

"Now for my first cup of coffee, but no, not yet. That roadrunner clacking its bill out in the bushes is just waiting for me to clear out so it can catch a baby cardinal for breakfast. I chunk a rock at him.

"After lunch, when I look out while cleaning (my furniture gets cleaned a lot near windows), I freeze in my tracks. There it is—a *big* hummingbird perched in the Chinese tallow! I almost come unglued, trying to think who to call *quick!* Nobody answers the phone, but finally I get Mary Anne McClendon.

" 'Put salt on its tail. I'll be right there,' she says.

"While I'm anxiously waiting, the bird comes to the feeder—and there's that clacking. I realize I've been throwing rocks at this beautiful stranger all morning.

"Mary Anne sees the bird and says, 'Call Fred Webster.' [Fred was the regional editor of *American Birds*.]

"When Fred Webster sees the bird, he says, 'Call Edgar Kincaid.' [Edgar was the editor of Oberholser's monumental *Bird Life of Texas* and also knew the birds of Mexico.]

"Edgar immediately makes me feel like the queen of birdwatchers by telling me I have in my yard a Green Violet-ear *(Colibri thalassinus)*—strangely enough, a bird that is not a migrant.

"Photographers from the Texas Parks and Wildlife Department take pictures of 'Big Boy.' Nancy McGowan, their artist, paints a portrait in watercolors; and thus the first authenticated record of this bird in the United States is clinched—right here at my own feeder.

"The news travels fast, far and wide, and birders come in droves, not only from Austin but from Dallas, Houston, San Antonio, Beaumont, La Porte, Baytown, Waco, and Wimberley.

"Fred Webster sends out a report to Joseph W. Taylor, who has 671 species on his AOU Life List. Taylor catches a plane from New York especially to gaze at Big Boy for an hour or so.

"In all, 112 bird fans came to admire my find, and for me that is 112 to add to my Life List of friends.

"My twenty-five wonderful days with Big Boy ended September 18, when he was last seen, and I certainly miss him and all the excitement he brought.

"James A. Tucker of Austin, founder of the American Birding Association, says best how I feel about my discovery: 'Happiness is being the most inexperienced birder and finding the rarest bird.'"

The irony is that while this exciting first-ever U.S. record was occurring only thirty miles from Adams Eden, Red and I were crawling through stickers in Arizona striving, unsuccessfully, to find the Five-striped Sparrow.

THOSE JEWELED JETS

Centuries before Europeans arrived in the Western Hemisphere, American Indians knew and marveled at hummingbirds. In their tribal languages they gave them names such as Shining Thornbill, Rain Bringer, and Magic Sky Bird. Some believed that hummingbirds were the spirits of the dead or that brave warriors killed in battle became hummingbirds. Others were certain it was the hummingbirds that brought rain.

The dried birds were used as ear pendants; their resplendent feathers were used to create pictures and, in some tribes, were reserved only for royalty. Everywhere, hummingbirds were considered good luck, and the birds' magic abilities were told and retold in many legends.

Today it is not fancy but established fact that causes us to marvel at the hummingbird. It is considered the most highly evolved of the nonpasserines, and it has become so specialized that it is placed in an order by itself, with its kinship sequence to other birds still undecided.

There is no fossil record of hummingbirds. They have evolved apparently in sequence with the evolution of the plants they feed on and pollinate and in competition with flying insects and other nectar eaters. If the journey toward the specialization of the hummingbird could be shown in time-lapse photography, what a movie it would make!

In flying abilities the hummingbird surpasses all other birds, and it depends on flight for all of its locomotion, no matter how short the distance. Hummingbirds seldom light on the ground, causing early naturalists to think they have no feet. Among birds, hummers have the largest breastbone in proportion to body size. In addition, the prominent keel and the flight muscles attached to it are proportionately larger than those of other birds, constituting from a quarter to a third of body weight in hummingbirds. Each feather on the bird's body is connected to a muscle, and especially those on the wings are sensitive to and compensate for changes in flight conditions. To turn very quickly or to back away from a hovering position, the bird spreads and turns its tail to lead its whole body in another direction, and there is an extraordinary coordination between the wings and the tail.

The bones in hummingbird wings are fused, so the wings move only at the shoulder, and this joint is so supple that it can rotate 180 degrees, giving the birds a sort of figure-eight flight pattern that gains thrust at both the upward and downward strokes. With this powerful muscle and bone design, hummingbirds can reach full speed instantly and stop just as abruptly. In take-off it uses no force from its feet; its wings do it all. Thus even at takeoff the bird is flying at almost full speed.

The extraordinary hummingbird can fly straight up or down, backward or forward, zigzag, and even upside down, and it can pivot on a fixed point. Hovering requires the most energy for wings must move unceasingly.

The maximum speed for one species, the Ruby-throated, is thirty-five miles an hour, but in battle or in a courtship dive the bird's speed might reach sixty miles per hour. However, hummingbirds cannot soar, for though they have ten stiff primary feathers as most other birds do, hummers have only six or seven stabilizing secondaries.

Despite their flying agility, hummingbirds can become a snack for anything from a hawk to a frog. Robert Duncan of San Antonio sent me the following account.

"While staying at the Peabodys' Mile Hi hummingbird lodge in Ramsey Canyon near Hereford, Arizona, my wife, Ruth, called me quickly to our second-story window," Robert wrote. "At the feeder just outside, a Wied's Flycatcher [now called Brown-crested] suddenly appeared and struck a male Broad-tailed Hummingbird and stunned it.

"The big-headed bird perched nearby in an evergreen and immediately began tearing and eating the smaller bird. I was told of four other recent similar occasions.

"A seven-inch bird capturing and eating a four-inch bird is less amazing when one considers how much of a hummer's length is in its slender bill."

Hummers are designed for life in the fast lane: their body temperature is higher and their heart and breathing rates are faster than those of other birds. They have relatively the largest brain of all birds (4.2 percent of body weight).

With the eight-and-half-inch Giant Hummingbird flying at about eight strokes per second and a three-inch Ruby-throated at about fifty-five strokes per second, and with each wing cycle lasting 1/500 of a second, fuel consumption for all hummers is extremely high. Percentage-wise, hummers require more energy than any other warm-blooded animal, and a daily intake of food is more than half the bird's weight. It is thought that the world's smallest warm-blooded creature—the Bee Hummingbird of Cuba, which weighs about one-fifteenth of an ounce—would not be able to eat enough to stay alive if it were any smaller.

For their size, hummingbirds have a large number of feathers, giving them excellent insulation, but the actual number of feathers is relatively small, ranging in number from 1,450 to 1,650; by way of comparison, a Brewer's Blackbird has 4,915 feathers, and an Eastern Meadowlark has 4,607. A few hummingbird species have special feather adornments such as crests, beards, horns, bibs, puffs, or long decorative tail feathers that may make a humming sound.

The feathers of most birds are colored by pigment, but the extraordinary scintillating colors of hummingbird feathers are the result of a process called interference, too complex for a full explanation here. Briefly, their wing barbules do not zip together, as in other birds, but instead are covered with stacks of microscopic elliptical plates that are made of melanin film and have layers of air trapped between them. The system both refracts and reflects light waves. However, all of a hummer's duller dark feathers are colored by pigments, as in other birds. The basic hummingbird color is green.

All hummingbird bills are slender, but they vary greatly in length and shape, from five-sixteenths of an inch to five inches long, and from straight to extremely curved, depending on the adaptation of a species to its favored blossoms. Usually, the tongue of a hummingbird divides into two fringed parts, and it is surprising that the birds do not suck nectar as if through a straw. Instead they lap nectar, and it is drawn up by capillary action.

Despite the shape of their bills, hummers can actively hawk insects in the air but usually probe them from flowers, tree bark, or spider webs. In the extremely rapid digestive system of hummingbirds, the remains of insects may be voided in only ten minutes after they are eaten. Almost 100 percent of the sugar that hummers eat is converted into energy.

Like other birds, hummingbirds have a higher body temperature than mammals, usually ranging from about 102 to 108 degrees whether at rest or active. In sleep, body temperature may drop as much as 8 degrees. To save energy, hummingbirds can go into a much deeper sleep called torpor, in which their temperature may fall as low as the surrounding air and their metabolism is reduced to a fifth of that of a hummer sleeping normally. If disturbed in torpor, they can chirp, but in case of danger they cannot rouse instantly from this short-term hibernation and must have a warm-up time before they can fly. Arousal may take from as little as ten to fifteen minutes or as long as an hour. This phenomenon, of course, is what led to the belief by both the American Indians and early naturalists that hummingbirds could die and then come back to life.

Since hummers occur from Tierra del Fuego to Alaska and Nova Scotia and from sea level to fifteen thousand feet, those in colder temperatures are forced to find shelter in caves or other protected areas at night or during storms. These species also build their nests in such retreats.

A resting hummingbird breathes 250 times per minute, whereas a pigeon breathes 20 to 30 times per minute. A hummingbird's heart beats about 500 times a minute but can reach a rate of 1,200 beats per minute in an excited bird.

Hummingbirds are not singers, with the very best song only a short little warble, but in dark, thick jungles males make use of song to attract females. In more open areas many species have elaborate mating flight rituals. The actual mating act is quite brief and is the only contribution males make to reproduction. However, males will fiercely defend their territory from an intruder, be it a butterfly, another bird, or a human.

Eggs are proportionately larger than those of other species, and the incubation period is long: eighteen to thirty-eight days. Nests are works of art. Held together with a plentiful supply of cobwebs, they are strong, very well insulated, and tastefully camouflaged with lichens. They can retain strength as they stretch with the growth of the young. The young are fed by regurgitation.

Attrition is high, but since hummingbirds have relatively long lives of about five to seven years, and since there may be a second or even a third nesting in a season, their numbers can be maintained. Individuals have lived twelve to sixteen years in captivity.

In his journal on October 21, 1492, Christopher Columbus wrote: "Little birds . . . so different from ours it is a marvel." From this first written mention, the exploration of the marvelous hummingbird by Europeans proceeded slowly and with a mishmash of misinformation passed along from one cabinet naturalist to another. Because of its size, the hummer was even dubbed the "bird-fly."

By 1630 the name "hum bird" was used in the English colonies, and it first appeared on a printed page in Captain John Smith's history of the colonies. Raphael, the great artist, apparently saw a specimen, for he painted a pair of hummingbirds on pilaster 11 of a loggia at the Vatican in 1519.

By the middle of the eighteenth century, hummingbird skins had become popular not only for display in glass cases on drawing-room tables but also for commercial use in jewelry, ladies' hats, dresses, crafts, and other decorative items. The result was the worst pillage of a wild creature known to that date, with the tiny beauties killed by the thousands to meet demand.

In one auction in 1888 in London, 12,000 skins were sold in a single day. Another dealer imported 400,000 dried skins in one year. It is probable that some species were eliminated entirely. Certainly there are skins remaining from those days that have never been matched to a living bird.

Thus exploitation by humans has been a major threat to hummingbirds, and the threat continues today with the use of pesticides and the loss of habitat due to agriculture and ranching. Horticulturists contributed a report of the following incident to my files.

Victoria Sudsbury of Rio Linda, California, kept hearing odd squeaks from the morning glories outside her kitchen window and found a female Anna's Hummingbird with a vine entwined around its neck. When she grasped the tendril, it turned out to be the arm of a praying mantis, a six-inch insect that had already nipped the bird's throat enough to bring blood and make feathers fly. When the mantis transferred its attention to Ms. Sudsbury, both bird and bug escaped to freedom. This giant mantis from the tropics had been imported by rose growers because it feeds on the aphids that infest roses.

Hummingbirds are the second-largest family of New World birds. When my column, Bird World, began publication in 1965, there were 319 species of hummingbirds recognized. In 1998 the American Ornithologists' Union and the American Birding Association recognized 345 species, plus or minus, in 109 genera. In Clements's *Birds of the World* for 2000, the number of hummingbird species was listed as 335.

Yes, hummingbirds do fly across the Gulf of Mexico. Fred Bosworth, in *The Last of the Curlews,* describes how northbound migrants gather in large numbers on the Yucatán tip of Mexico before braving the five-hundred-mile Gulf of Mexico crossing: "For many of the smaller songsters such as Bobolinks, thrushes and warblers . . . it was the migration's most rigorous ordeal and the time of starting had to be carefully appraised. By mid-afternoon many were not returning from their test flights. . . . Many hummingbirds, hardy midgets weighing no more than a tenth of an ounce, had started with the others. But now they were far ahead, outdistancing them all, their tiny wings churning the air with seventy beats a second. Most of the birds would fly twenty hours before they reached the American mainland. The curlews would take ten hours. The hummingbirds would do it in eight."

Now wind tunnel studies made at the University of Texas in Austin by Peng Chai and his co-researcher, Dmitry Grodnitsky of the Russian Academy of Sciences, have revealed that the maximum average flying speed of the Ruby-throated Hummingbird is thirty-five miles per hour. Amazingly, this matches the flight speeds of birds of much larger size, such as crows, magpies, and gulls. These studies of the hummers' maximum flight performance were used as a yardstick to measure all its other flight behaviors.

Grodnitsky used a harmless, low-power laser to illuminate dust particles in the air, disclosing the wing motions of the birds as they flew in a wind tunnel. Yes, even in a scientific study their wings make the usual audible hum.

Chai, following federal and state guidelines, captured and held eight male and female hummers, which he placed, two to a cage, in his laboratory near a window at the University of Texas Department of Zoology laboratory.

"How do they behave?" I asked.

"They're *very* noisy. They have such a strong instinct to dominate that they will fuss at all the birds in adjoining cages. We try to eliminate quarrels between the two in a cage by giving them separate syringes to feed from, but sometimes one wants both syringes, so we have to move that one to another cage."

The researchers placed the birds and a feeder in a Plexiglas box filled with helium and oxygen in a density only one third that of normal air. This construction acted as a sort of treadmill for researching the maximum flight performance of the hummingbirds. Their oxygen consumption was discovered to be one of the highest ever found for a vertebrate and about ten times higher than that of humans.

The birds were well fed on a commercial food especially designed and balanced for hummers, and they also could catch fruit flies that flew out of small bottles. With a perfect diet and much exercise the birds were in excellent condition when they were released in time for migration in September 1997. However, they were not banded, and it was not known how well they survived.

Hummingbirds have attracted the attention of the military. The U.S. Navy has been interested in their quick takeoffs and landings, which could be adapted to aircraft carriers. There is also great interest in what is known as micro air vehicles (MAVs), a new type of very small airplane suitable to fit inside a soldier's backpack. These vehicles must be no larger than six inches, weigh from two to four ounces, have a minimum range of six miles, a cruise speed of twenty to forty miles an hour, a mission duration of twenty minutes to two hours, and a payload of one ounce.

Equipped with microcameras, sensors to detect toxins, and other technologies, MAVs could act as communication relays and fly in complex and constrained locations such as a burning building. Undoubtedly, they would also find civilian applications. Not only would a micro air vehicle be small, but every component of it would also have to serve more than one purpose. Thus in 1997 Tifenn M. Boisard, an exchange student from France studying mechanical engineering at the University of Texas, began investigations of the Ruby-throated Hummingbird to ascertain if the basics of its flight behavior could be adapted into the design of a micro air vehicle.

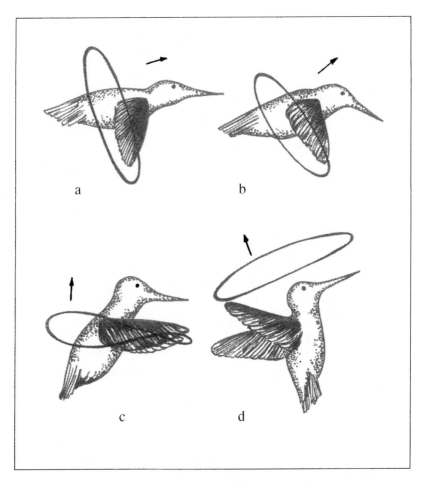

Hummingbird wingtip paths during several flight phases: a, *forward 26 miles per hour (top speed);* b, *forward 4.6 miles per hour;* c, *hovering;* d, *backward flight. (Drawings by Nicole Bermudez)*

Although the hummingbird is one of the strongest natural flyers, its method of flying is one of the least efficient. Peng Chai found that large hummingbirds could lift twice their body weight in short burst performances, but smaller ones could lift only half their weight. Boisard, in her conclusion, stated it would be extremely difficult to create an aerial robot that performs as effectively as a hummingbird, the result of thousands of years of evolution, adaptation, and perfection.

***

It was November in Massachusetts, and a small crew of rescuers was working desperately to save a very small life. Nets and food had been to no avail, and finally Tom Gagnon gave up and drove home, where word came that the tiny hummingbird had finally flown into a garage.

Gagnon returned and, using a toothpick, released the feet of the hummingbird from an old fan housing where it had lit in exhaustion. He carried it home to Florence in a pillowcase.

When released, the bird didn't fly, so Gagnon used a Q-tip to put a little bit of sugar water on in its bill. After two sips, the bird flew around in Gagnon's greenhouse.

According to natural law as well as international law, this bird should have been sipping nectar in Mexico, but instead its migrating instinct failed, and it had stayed at Lucy Fieldstad's feeder in Agawam, heedless of the impending winter. Finally, its hosts realized that the bird was not going to migrate, and through a series of connections the situation was presented to the U.S. Fish and Wildlife Service, which issued a capture permit.

So this is the story of how what finally became a full-grown Rufous Hummingbird spent the winter and early spring in a greenhouse in Florence, Massachusetts.

But the story is not over. In 1997 the bird showed up at the same feeder, remained late into the season, and then was recaptured and placed in the same greenhouse for a second winter.

Wayne Petersen, vice president of the American Birding Association, said a dilemma such as this raises several ethical questions. Should people leave feeders out after most hummers have left? Is it appropriate to capture and restrain a wild bird to keep it from perishing? Is it ethical to capture a bird for the purpose of positively identifying it? His advice: Take feeders down at a logical fall date. Birds then have no choice but to migrate. (This is Petersen's advice—wrong or right. Some others say that keeping feeders up doesn't affect a bird's urge to migrate. The Texas Parks and Wildlife Department, for example, states on its Web site: "Leaving hummingbird feeders up will not stall [the birds'] migration, but may help some of the stragglers along who were slowed by illness or late nesting attempts.")

And now, back to Peng Chai. He noted, "In nature hummingbirds get sugar from plants as a reward for their pollination service. Now sugar water provided by humans from thousands of artificial feeders attracts many hummingbirds as an easy source of food. How this human interference potentially deprives the birds of their useful role as pollinators in nature remains unclear."

In connection with this thought here is another anecdote from my files. Blooming fruit trees remind me of the Gene Lloyds of Austin and their hummingbird feeders. Most of the Lloyds' neighbors had fruit trees, but the trees seldom bore much fruit. When bees began to come to the Lloyds' five hummingbird feeders, they decided to take them down for several days. Both bees and hummingbirds stayed busy instead in the fruit blossoms, and that year the neighbors got a good crop of fruit.

This incident poses a further question: How much vital pollination of the plants of the natural world is "stolen" by hummingbird feeders supplied by humans?

CONVERSATION(?) WITH A ROBIN

The yard around our recently acquired, almost fifty-year-old house, has been long neglected and junked up, and as I dig new flower beds and garden spots, I find broken toys and pieces of metal and faded glass, along with countless bones buried by dogs long since gone to their happy hunting grounds. Among the chards I also uncover many choice living tidbits coveted by birds, and all through the winter I've enjoyed the company of a robin each time I worked.

The American Robin is one of our country's most adaptable birds and is common in most of the United States, living in such close association with humans that it seems like a member of the family. Often the cheerfully voiced redbreasts seem to reciprocate the feeling by being very tame in parks and yards. My robin was no exception. Usually it remained less than twenty feet away, watching me slave and calling out from time to time in a half-flat robin note to remind me it was waiting. We humans must be cautious in interpreting wild animals' actions, but it seemed my robin's notes became peevish and complaining when I didn't move away promptly to let it clean up the bounty I'd uncovered. As soon as I went to another part of the yard, it would scurry to the spot I'd just dug, brushing the soil away vigorously with side strokes of its bill.

The closest it came was about four feet, where it sat in an arborvitae, impatiently watching me struggle with a worn-out rosebush that must have been there for centuries. When I finally got the rascal out and triumphantly moved off with it, my dinner guest's usual flat chirps became low, musical, and rapid, seeming to congratulate the both of us as it gobbled its meal.

Recently, as I dug the last of about thirty feet of an asparagus bed, it stood about ten feet away, and this time I knew Red Robin was quarreling at me, so I looked straight into the bird's white-circled eye and told it in my friendliest voice, "Just go ahead and get what you want."

As I continued to talk in a low tone, the bird moved in closer, picked up a

June bug larva, took its time to worry and soften it to the right texture, and then downed it before sedately moving back to its usual "safe" distance.

By moving slowly and voicing low and monotonous assurances, I've "talked" myself around jackrabbits that were in my path (as close as four feet), had birds bathe while I talked persuasively from a chair an arm's length from the birdbath, and even talked Greedy, one of our deer neighbors, into taking bread from my hand. These wildlife friends had grown accustomed to me, their unaggressive neighbor, over a period of time.

I haven't been able to resist giving bird names to all the members of my family, and this custom has now extended to the fourth generation. My oldest great-granddaughter is our Robin, so it was natural for us to arrange a trip for her to a robin roost before sunset. Clifford Shackelford of Partners in Flight said the roost had comprised several million birds, but it was much smaller now. The chattering birds were coming from all directions to a juniper-oak forest, but it became obvious, gauging by the sound, that we were not at the center of the roost. The flight continued till dark.

It was a good experience with our young "thrush," but far from the one Red and I had witnessed more than two decades ago. We pulled off a highway west of Austin to watch a stream of robins two to three hundred feet wide, flying from horizon to horizon. The spectacle lasted so long, with no sign of diminishing, that we finally got back in the car and drove home.

My dad told me that when he was a child in North Texas, his family sometimes dined on robins. Dad would have been 115 years old today. Texans' tastes and values have changed a lot in not much more than a lifetime.

In the winter, when food is scarce, robins sometimes eat the berries of the chinaberry tree, which by that time may be fermented. On one of his visits to Texas, Roger Tory Peterson saw robins losing their balance and skittering around and was astonished when he was told that the birds were drunk.

DINING ON THE WIND

Alone in a motel room in a strange city was not the way I had planned to spend this day. The phone was already dead, and the television showed trees whipping and bending and a sign tumbling down a street. The announcer's words were, "The wind is now in excess of ninety miles per hour, and increasing rap—"

The lights went out the same instant.

I sat on the bed, listening to the roar outside and feeling furious with it, then helpless against its force, and finally I stared out into its fury, realizing fully the danger of being its prisoner.

"Change the subject," I ordered my mind. The window gave enough light to read by, but I had no magazine or newspaper, not even a Gideon, and only two field guides, both of which I had long since almost memorized.

"Well, you have your notepad," I suggested. "Find out what kind of company you make." I scribbled a word or two, then went from dull to blank. The wind had taken charge, and its constant howl had blown away my thoughts.

The wind. I was studying it like an attacking enemy and realized it was blowing slightly *away* from the window. It appeared that the corner of this building was providing a partial windbreak. I began breathing the hope that the storm won't crash its way inside.

The subtropical Lower Rio Grande Valley of Texas abounds in flowers, but I never expected the traditional Christmas beribboned potted plant to become a hedge. Outside the window, poinsettias had grown into a four-foot hedge, a luxuriant wall of flaming flowers all desperately fighting with the wind and rain.

Then I saw the baby.

"Poor innocent," I said to it through the window. "This is not a good day to be a baby mockingbird too young to fly. Too bad the hurricane is going to be your flight instructor."

I expected to see the fledgling make a quick takeoff, but, no, the building

was giving Baby enough protection for it to cling awkwardly to a low poinsettia stem. I could see its bill opening wide again and again with a feed-me
order to Mama Mockingbird.

Surely she couldn't hear the demands above the roar, yet she was obey-
ing the calls, and her feeding task was greatly simplified by powerful wind
gusts that dislodged choice bugs from their hiding places and delivered them
to her.

This ideal dinner scene began to change as the gale increased. Insects
gained such flight speed that she scarcely could race them. A return from
downwind became an engineering problem.

"Come on, Mama," I found myself coaching. "Don't give up. Come on,
you can do it. Just bank a little bit more on the approach."

Baby's perch took a downward slant as wind and rain grew more violent.
If the hedge went, Baby would go too.

I found myself talking to the hedge: "Forget that to most of us you're just
a potted plant. You're outdoors now and living the life Nature intended for
you. Come on, be a champion. Dig in with your toenails or whatever you've
got down there in the mud and hang on."

My coaching had little influence as Baby's particular poinsettia slowly gave
to the relentless, constant power of the hurricane. The once-glorious plant
was almost prostrate now but miraculously still gave vital inches of protection
to its young tenant, who, artlessly unknowing, kept announcing its appetite
soundlessly into the tempest.

"Mama, how can you keep this up?" I marveled. Had the wind slightly
changed direction? Her drenched feathers made dodging the leaves and
branches that filled the air an aerial feat, and a large section of a tree somehow sailed and tumbled by out there without harming her or destroying what
remained of the hedge.

Now her expeditions resulted consistently in somersaults, and the wind
turned her feathers inside out, but somehow each time she managed to right
herself and catch another tidbit, with which she managed to struggle to Baby,
mightily bracing herself long enough to target her infant's mouth.

Oh, no! I slumped in my chair. The wind had won. Baby had been unceremoniously dumped to the ground. But, hey, look! There's a small hump
of mud there, and Baby has a toehold on it above the surrounding water.
And here comes that miracle of a Mama, digging her claws into the ground,
wading the water, and pecking hard at what swept past so that she could stuff
the essential nourishment into her little glutton's throat.

Night was approaching as the wind finally blew itself out and left the stage.
The eight-hour struggle had ended, with Mama and Baby the winners.

Their cheerleader had battled every inch of the way with them and was as worn out and wrung out as they were, but, unlike Baby, she was very, very hungry.

But, wow! I realized that it was crazy, but I was feeling absolutely glorious. It must be because I had just witnessed firsthand one of the little miracles of the natural world.

We never know when we'll see the next one.

OUTGUESSING THE WEATHER

In October 1967 there was some word about a mild disturbance in the Caribbean, but with our work schedule incomplete and the forecast for the storm many miles east of us, the Adams pair decided we'd be safe 130 miles inland at Falcon State Park where, in the far distance across the huge lake, we could see Mexico.

We laid in groceries and gasoline, and Red hitched the trailer to the car for a fast getaway if necessary. He even ran a light rope from a small tree and secured it to the trailer. Thus battened down, we continued to enjoy the Black-throated Sparrows, Curve-billed Thrashers, and Cactus Wrens, which without weather warnings continued to sing and gargle at our door.

The coyotes were still merrily yapping as a storm struck in the dark morning hours, and though the wind was fierce and the rain drenching, it didn't seem anything to worry about, for the blackbirds came off their roost only a half hour late, the Osprey worked for fish in the face of the gale, and the Scaled Quail still ventured to pick up our scattered grain.

Then the storm became Hurricane Beulah.

The Road Roost was parked with a sweeping view of Falcon Lake, created by damming the once-mighty Rio Grande. The huge body of water was now a frothing mass of whitecaps topping waves ten to twelve feet high, and Red and Marjorie and everything else hereabouts took cover. We learned later that Hurricane Beulah had come ashore as predicted but unexpectedly encountered a strong front that turned the huge storm westward.

I couldn't guess how many times during that anxious day and night Red and I exchanged searching gazes as sudden gusts made us wonder just how much weather a twenty-two-foot travel trailer can withstand. With wind gauges broken, the consensus of the professionals was a minimum of seventy-five miles per hour. The rain gauges totaled twenty-two inches, but did that include all of the rain that flew horizontal? This was one occasion when Red was happy about all my heavy books and files.

We were fearful the trailer windows would blow in, but they were the wind-out type that lock in place, and only one of them leaked. At nightfall, realizing my helplessness against nature's fury, I lay down to sleep as best I could.

Red, guardian of the family, stayed awake much of the long night, and at times when the trailer swayed unsteadily, we would talk to each other as calmly as we could, sometimes about our children and their lives. And still the wind blew.

At last even Hurricane Beulah abated, and as soon as the wind let up, Red unlashed the rope and unhitched the car for an inspection tour. It seemed everything in the wild world, from birds to lizards, was out looking for break-fast. Then—to our astonishment—Beulah was back with a vengeance and blowing the reverse direction! The lull we had experienced was her deceiving eye. It was only later that we learned it had centered south of Zapata and passed directly over us.

Unsuspecting, we were caught outdoors as the eye passed, and we realized we were truly lucky that Red was able to apply the force needed to open the trailer door just in time for us to get inside and endure another eighteen hours of it.

In weather like this, what happens to wildlife? As expected, we saw every critter fighting any way it could to survive. There were short lulls when move-ment, though difficult, was possible.

I saw Scaled Quail, Curve-billed Thrashers, and a Golden-fronted Wood-pecker eat the fruit of a chittimwood shrub just outside our door, probably an emergency food for all of them. Barn, Cliff, and Rough-winged Swallows and a couple of late Purple Martins, Scissor-tailed Flycatchers, and an Osprey fought the gale for food in any lull they dared to risk.

We often heard coyotes' howls above the wind's howls and felt they weren't going entirely hungry, for we saw a bedraggled rabbit nibbling a wet meal under protecting brush.

At last the storm was passing, and as soon as the wind would let us, we were out making a survey. Some highly unexpected visitors had blown to the huge freshwater lake. There were Royal, Sandwich, and Gull-billed Terns, being harassed by two of those fierce seagoing pirates, the Pomarine Jaegers. It was an exciting experience to observe these swift birds as we stood on solid land instead of a lurching ship.

There were hundreds of the more common Black and Least Terns, at least a dozen adult and juvenile Black Skimmers, many Laughing Gulls, a few Snowy Plovers, and many flights of ducks, none close enough to identify except Blue-winged Teal.

Other birds far from home were those giants with the seven-foot wing

Silhouettes of jaegers, frigatebirds, and terns.

span, three Magnificent Frigatebirds, peculiarly in what seemed a playful mood with each other. Our major prize, a Lifer for both of us, was two striking black-and-white Sooty Terns, birds rarely seen for an ABA list except by taking a trip to the Dry Tortugas Islands off the Florida coast.

If anything good can be said for hurricanes, it is that they make birding extraordinary. As for predicting Texas weather—don't.

***

An addendum: Sandy Sprunt, noted bird scientist of Florida, passed along this statement: "Actually, the secondary kill from hurricanes can be much more serious for birds than the winds themselves. For instance, Hurricane Inez didn't do much damage per se, but the severe winds blew all the little insects that migrating warblers depend on out of the trees, so that the birds' food was basically destroyed for days. Hunger made them lose all fear, and they were out on people's lawns and on every open space trying to find something to eat, even lighting on people's shoulders and flying into houses."

TURKEY WILD, TURKEY TAME

I used to know quite a bit about Wild Turkeys because Red and I some years ago began making a movie about these truly American birds. It was to be a simple film for school kids until our son, Lew, saw some of Red's 16 mm footage. We had miles of it, most of it high-quality, and Lew was seeing it with Hollywood eyes.

Red had found a trio of toms in fierce competition for the females. As they strutted in full pride, their powerful wings stroking the ground, wattles swollen blood red, resplendent plumage glistening in metallic gold, bronze, and copper, and richly banded tails fanned to flag attention, they often sent a scintillating shiver throughout their glory each to prove he was the Chosen One. They were nothing less than total magnificence.

"That's it," Lew decided. "We're going for the big time."

Lew had started running cameras at NBC at age twenty, had worked with many network shows and producers, and now had become a producer heading his own company, so we sat at attention and listened respectfully.

"We're going to discover the world's first talking turkey," he announced— we scarcely blinked—"and it's going to be America's great good luck that the language this bird has mastered is English!"

This miracle of Nature would then be our star in a comedy special, which also would include a passel of Red's wild turkey footage.

Our script called for a haughty Wild Turkey to ride in a limousine through a cheering crowd to arrive—*talking*—at a major television studio, where he would be received amid great ceremony. This majestic hero would then be escorted to the makeup studio to prepare for his first worldwide television appearance.

There would then occur such a large-scale flurry of activity by a swarm of renowned makeup artists that our star would be completely hidden from view. This story line was conceived before the advent of computer magic, yet there seemed to be no limit to what the glitter world could produce.

The bird that emerged from the hands of these makeup geniuses would be none other than Jonathan Winters or a similar great-name comedian dressed in finest feather as a magnificent and imperious King Turkey. This commanding monarch would then take total control of the program.

King Turkey had come with an extraordinary mission: he wanted to prove that the Wild Turkey had saved America. The goal of this bird-monarch was to have the world's most illustrious artists create a memorial that would be erected near the Lincoln Memorial. Its message: If it hadn't been for the plenitude of Wild Turkeys, the colonists would never have survived. (It is a fact in America's history that another name for turkey breast was often "turkey bread.") As well as I can remember, King Turkey also wanted schools established to teach turkeys how to speak English.

Of course, all this drama would be so cleverly fantasized by Hollywood's best writers that it would be totally, totally real and everybody would simply die laughing.

So here we find real-life Marjorie and Red, driving out into the desert to talk in detail with a wildlife trainer. This trainer's home base had many cages, most of them occupied with wild and woolly animals who were expected at some point in their lives to earn their keep by appearing in front of a camera.

When we arrived, the trainer was busy teaching about a dozen gray wolves to follow his orders. They did. Thus when it came our turn, and he swore to us that he could coax a turkey to ride in a limousine and walk proudly into a big television studio, we almost fell at his feet. In today's high-tech world all of this could be done using computer-generated animation, but this was in the dark ages of the 1970s.

Somewhere in here we needed to get footage of the birth of a turkey, so Red and Marj procured a half-dozen Wild Turkey eggs and kept them warm and properly turned for almost a month. Sure 'nuf, we became proud parents of some fluffy, cute, and active baby turkeys.

However, as they cheeped here and there, we discovered a disturbing fact: they weren't eating anything. We made all kinds of offerings without success. Quick action was needed—our babies were going to starve.

We sought advice from the state wildlife service, from hunters and poultry breeders, and anyone, but it was our neighborly feed-store man who finally set us straight.

"You didn't know baby turkeys don't know how to eat?" he asked in a highly accusatory tone. "You gotta teach 'em," he ordered.

Duh?

So that is how we began calling all our baby turkeys Dumb-dumb, except for one that we called Double Dumb-dumb. So here is Mama Turkey, otherwise known as Marjorie, on her knees talking to her poult babies, you know, a *put-put* here and a *put-put* there and a big scratch here and a little scratch there with a twig in the dirt, and lo! Here comes a piece of hard-boiled egg. It didn't work. Finally, Red and I began putting dampened turkey feed in their bills. Alas, they let it fall in our laps, but we persevered. Finally one gulped it down and recognized there was more of the same thing in a dish, and the others followed suit. Finally they began scratching outdoors on their own.

About this time we visited our farming friends Herbert and Verlene Bohls in Pflugerville, Texas, and we drove through a field alive with grasshoppers.

"Stop the car," I ordered. "I've got to catch bugs for my poults."

Pflugerville folks are right neighborly, so it ended up that the four of us had a jolly time flouncing and dancing and falling on our knees and proving that grasshoppers truly can hop and fly and dodge and slip out of your fingers. We remained friends, and to this day we all remember the event as if we are still kids.

It was amazing how intelligent the Marjorie Turkey became. She quickly learned her babies were not carbon copies, but individuals with separate personalities. King Tom had a diamond shape on his head. His constant understudy was Furrowed Brow. The one they both pecked the most was Squarehead. There was still Dumb-dumb and Double Dumb-dumb.

Since Mama Turkey Marjorie was the first living thing they had seen when they pecked their way out of their eggs, they imprinted on her, and, like little wind-up toys, they mechanically followed her wherever she went. Think of the responsibility thrust on the human! All these precious little lives were totally dependent on her.

In only four days she faced a major problem. The little tom turkeys' downy feathers had scarcely dried when each one of them realized they were born to compete, and each one of them declared total war on the others.

With his tiny baby wings flaunting power, one would approach a rival and grasp its neck with his own. Entwined, the two would battle by pushing and turning to force the other into submission. It was a spectacle not to be imagined—baby turkeys trying to wring each other's necks. Wow! We could see this really was perfect for prime-time TV.

Fortunately, Nature does allow one combatant to finally win the title of King Turkey, and no one dies.

Yet Al Springs of the Texas Parks and Wildlife Department told me, "I was out in the woods in the Streetman area when I came upon two big gobblers in combat. Their necks were so entwined, I could scarcely tell them apart. Apparently, their battle had been going on for a long time, and as I walked toward them, they didn't move. When I came within a few feet. they still didn't move. They had battled so long, they had come very near to actually wringing each other's necks."

Of course, we wanted footage of our little gobblers' battles, but when Red set up the camera, not one turkey would challenge King Tom. The solution was to unceremoniously declare time-out and remove King from the group. Immediately Tom Two began fighting Tom Three, and we got all the movies we needed.

Our healthy turkeys soon grew too large for a cage, and with all the millions of wild domestic cats that roam America's backyards, they certainly were not safe outdoors, so we carted them off to a small turkey farm.

When we turned them loose in a pen of turkeys about their same size and age, our King Tom quickly was neck to neck with their King Tom. That was not the reception we expected, so we stayed to witness the outcome.

The struggle for predominance is so inbred that the conflict continued quite a while, but finally our King Tom won, and we left knowing he had found a happy home. All our turkeys had, except Squarehead. Everybody was pecking him.

It took time, sweat, research, and even some money, but we got a complete history of the turkey on film from birth to death. We even had a basic script for our Famous Turkey Comedian.

The factor we hadn't taken into account was that at that time in Hollywood "turkey" was slang for a flop show. Undoubtedly, that is the sole reason this worthy epic never flew.

The theme of the show was the important and almost unbelievable truth about turkeys: this nation came very near to losing this wonderful species. Though there were probably as many as ten million turkeys in the United States when the Europeans arrived, this multitude of turkeys had been extirpated from most of the East and much of the West by the 1900s.

Somehow we superbright Americans took for granted (just as we do today) that there would always be forests aplenty, with all their accouterments, such as turkeys aplenty. Thus we killed the birds wantonly, then we killed them defensively when they damaged our crops, and finally, unknowingly, killed them when we cleared, burned, chopped, and plowed the homeland where they had to make a living.

To their everlasting credit, our nation's sportsmen-hunters were the first to demand protection for the Wild Turkey. It has been a many-year task for both government and private agencies to bring the turkey back. The triumph is that the turkey not only has been restored to the states where it historically occurred but also has been introduced to states such as Wyoming and Utah and even Hawaii, where it never occurred at all. Today Alaska is the sole turkeyless state.

In the southwestern United States the Wild Turkey was domesticated at least by A.D. 700–900 by Native Americans and, in some tribes, was considered sacred. In Mexico and Central America it was used as a sacrificial animal in bloody religious rites. Turkeys were an important trade item. Turkeys were portrayed in the art of many cultures, and some tribes mimicked the turkey in their dances.

The bird has continued to be an important and growing segment of our economy, and thankfully, it also continues as an intriguing wild creature.

The first thing to understand about Wild Turkeys is that they are athletes. They are highly adaptable and so hardy that bad weather seldom kills them. The exception is baby turkeys in the rain. They are apt to raise their heads into the rain, open their bills, and drown. Turkeys can survive snowstorms by staying in trees for several days; then their long legs and sturdy feet enable them to scratch through snow as much as twelve inches deep to find food.

With long, strong legs, a turkey may range as much as fifteen miles in a day's feeding, and during breeding season a gobbler may range over as much as twenty-five thousand acres. This is a bird that can run from twelve to fifteen miles an hour and up to thirty miles per hour in short spurts, and this running ability was one of the first traits to impress the early colonists. The flight speed of the turkey has been tallied from thirty to as high as sixty miles an hour.

The turkey has one of the strongest muscles known, its gizzard. This organ, smaller than a tennis ball, can exert a force of as much as eighty pounds. No wonder it can crack hickory nuts.

Back in 1975 I asked Caleb Glazener, the turkey expert at the Rob and Bessie Welder Wildlife Refuge near Sinton, Texas, "How intelligent is the Wild Turkey in comparison with other birds?"

Caleb laughed. "The turkey can be wary and smart beyond imagination, and then he'll do some of the dumbest things imaginable."

"In other words he behaves like humans?"

Caleb nodded. "Yes, we do have some characteristics in common," he said. "The turkey is not as wary as some of our waterfowl, which are extremely shy, and his behavior varies with the habitat in which he ranges and even with the different subspecies. But when their population numbers go down, they tend to retreat to the woods and become very spooky."

Students of this species agree in general that turkeys have an innate curiosity and are apt to gobble at sudden noises (an old tom in a flock was known

Red Adams, coming out of Roy Creek Canyon with a Wild Turkey. (Courtesy of Lew Adams)

to gobble when a jet plane flew over). They have a proverbial keenness of sight, are highly pugnacious, and are even playful, sometimes with a group leaping, skipping, and waltzing about seemingly just for the joy of it.

I quote here a hunter of 1880 who said, "The turkey is the wildest and the tamest, the most cunning and wary, the most stupid and foolish of all birds."

And I repeat here what has been generally agreed: "No man can say a mean thing with a mouth full of turkey."

185

Let's end this with a turkey revenge. It happened in Belize several years ago to Drew Thate. I caught up with Drew by phone in Montana, where he was currently a fishing guide.

"Well, a good story sure can travel," he said with a laugh. "Here's the way it happened. Are you familiar with Chan Chich Lodge in Belize?"

"Yes," I replied. "Red and I traveled there in our motor home. It's an amazing resort for birders in the middle of the jungle."

"And all the comforts of home," Drew said. "I was working as a tour guide for birders there, and that day I was taking a small busload of about a dozen people out to Laguna Seca."

"I remember it like it was yesterday," I said. "There are crocodiles in the lake there. And a little colony of oropendolas. Actually, we spent the night at Laguna in our motor home. I was in heaven."

"You'll remember the road is an old logging road and favored by Ocellated Turkeys?" Drew asked.

"We saw our very first ones there. It was like a dream—they're so beautiful."

"Well, a male was using the road as his display territory," Drew continued, "and as our bus drew up, he refused to move. Maybe the fact that the bus was blue made him want so badly to fight. Anyhow, I finally I got out and began to shoo him, waving my arms, but he firmly stood his ground. This went on for some time, so I turned my head to get back on the bus.

"That was the mistake. The bird instantly attacked me, grabbing my right thigh with both feet and sinking his talons right through my trousers. At the same time he was beating me in the chest with his wings and pecking toward my face. It was sort of like a one-bird cyclone hit me.

"When I got in the bus, I was good and bloody. In the tropics even a small scratch can get infected, and all the birders insisted I should get some medical care right away. Back at Chan Chich, the Hardings took over. Josie is a nurse, and Tom was a medic in the army. They figured the turkey's talons had gone as much as an inch deep, so they gave me a tetanus shot. That's one battle a bird won."

PLAYFUL RAVENS?

Do birds have a sense of humor? Does this incident witnessed by Pat (Mrs. Carl) Snider of Los Alamos, New Mexico, indicate they do?

"Two ravens swooped down on our cat fore and aft," Pat recounted, "and one grabbed the cat's tail in its bill. The cat swung around to slap it, but as it did, the other raven nipped its tail fiercely. When the cat turned back on him, the first raven seized his tail again. This alternating game of fun went on a long time before the cat escaped."

Pat also described another behavior she saw and heard: "As I walked out of a building in Los Alamos, I heard a loud *cra-a-a-ak,* and on a telephone poll beside the entrance was a Common Raven. When someone else came out the door, the raven immediately let out another croak. As I watched for at least ten minutes, every time anybody came through the door, the bird let out a loud squawk. I don't recall that the door itself made any sound when it was opened, so don't think this was any kind of conversation."

In Colorado, Red and I saw four Common Ravens roistering in and sliding down a steep snowbank. They flew back to the top and did this again and then again, obviously getting pleasure out of it. Were they tobogganing just for fun, or might they also have been doing this to clean and rearrange their feathers?

What about numerous other observations when ravens seemed to be playing by dropping sticks, then swooping down to catch them again and again before they hit the ground? Ravens have also been seen passing objects back and forth in flight to each other with playful diving, tumbling, and twisting.

Margaret Millar, a California mystery writer, told the story of Melanie, a raven that was raised by humans and resided at the Santa Barbara Museum of Natural History, where the bird's apparent lifework was overseeing the outdoor lunch area. Melanie not only appreciated hot dogs and potato chips but also was notorious for snatching any bright objects such as barrettes, bracelets, bright buttons, and earrings.

When Millar dropped her purse and her car keys spilled out of it, the raven promptly got the keys in her bill and flew off to the hillside, where she sailed around and around with them, then dropped them in the brush.

Since the hillside was thick with poison oak, stickers and possibly snakes, Millar envisioned herself walking home. Remarkably, after flying around awhile, Melanie flew back to the hillside, came up with the keys, and sailed back to the lunch table, where she gave up the keys in exchange for a hot dog.

Ravens have figured widely in folklore and legend, and modern science has proved they have mental powers of abstraction and memory that put them on a par with many mammals.

These birds have remarkable survivorship abilities, which make it possible for them to live even in arctic conditions. In those regions their feet have adapted with pads six times as thick as those of ravens in subtropical climates. These "corns" contain keratin, a poor heat (and cold) conductor.

Ravens figure in many legends and have been revered in many countries. For instance, I have been told that in the country of Bhutan the raven is considered such a sacred bird that to kill one is a crime as serious as killing one hundred monks.

"Oh, yes," Rose Ann Rowlett agreed during a recent telephone conversation from her home in Arizona. "The raven is the national bird of this beautiful little country. I'll be the tour guide there for a group of birders next week. Because of the moisture that blows in from the Bengal Sea, the tree line in the Bhutan mountains is one of the highest in the world, at about thirteen thousand feet. Bhutan ravens are an arctic type that live only in the mountain highlands." I find myself wondering what kind of slapstick humor is inspired by such dizzying heights.

I didn't find much humor in the legend that ravens have been living in the Tower of London for a thousand years. The legend also has it that if the ravens disappear, England will be no more. That should keep the ravens safe and sane for the foreseeable future.

And let's not forget that "at the end of forty days Noah opened the window of the ark which he had made, and sent forth a raven, and it went to and fro until the waters were dried up from the earth."

And that was no laughing matter.

BATTLE OF A LIFETIME

I still remember the awe I felt. On the surface what we saw was a simple thing: two birds flying around in a circle again and again, yet the sight got not only my head spinning but my mind too.

Red and I knew there were at least three Golden-cheeked Warbler males singing for territory in this part of the canyon of Roy Creek, and apparently we had reached this one's realm. He was singing vehemently; then suddenly in a rush of wings he was attacked by another male.

It was obvious from the pattern of flight that we were standing in the real estate that male Number One considered the heart of his territory, for repeatedly he chased his challenger in an almost identical pattern around the same spot.

It was not until perhaps twenty such circles had been swiftly maneuvered that Red began to count, "One," and I joined in. Our count reached the thirties, then the forties and the fifties, with the birds several times going momentarily out of sight behind taller trees but always quickly coming back to the central clump of large trees.

Warren Pulich's historic study of the Golden-cheeked Warbler showed warbler territories to average a little more than four acres. This territory seemed to be much smaller (or at least the battleground was), so this "grocery store" must be especially rich in the necessities this endangered bird needed to raise a family: Ashe juniper ("cedar") trees old enough to shed strips of bark suitable for weaving the beautifully crafted three-inch nest; the plateau live oak, host to the caterpillar webs used to hold the nest together; and the Texas (Spanish) oak that hosts the green caterpillars the adults favor for feeding their young. A bonus was the sparkling creek water for plentiful bathing.

As the contesting birds continued their dizzying performance, Red and I were also twisting heads and turning bodies to keep up with their fiercely combative speed. The warriors were oblivious of us, and clearly their every action was stemming from their innermost beings.

Two male Golden-cheeked Warblers battle it out. (Illustration by Nancy McGowan)

At this frantic pace we soon began to feel that each circle would be the last, and at circle fifty-five, male Number Two paused (hopefully?) for perhaps a count of two, but, no, Number One hurled himself valiantly at him and they were off again. We were now at circle sixty, and this time feathers flew, sailing and circling gracefully on the slight breeze.

There was another pause, and then at circle sixty-two the exhausted intruder disappeared from view.

We guessed that the contest was already under way when we arrived, and we didn't start counting circles until we had seen about twenty passes overhead, so the total number of circles the birds made was at least eighty . . . maybe more.

The Golden-cheeked Warbler is about five inches long, and the male weighs 0.36 ounce. At 0.33 ounce each, three females could be mailed for a first-class postage stamp. How could so much valor come from such a tiny warm-blooded entity?

Pulich states that the longest conflict he observed between males lasted only three minutes and most lasted not more than a minute. Surely this one we witnessed was a record and deserved special note.

Records always have their statistics, so I began scheming comparisons. What could be a comparable human feat? And what could such a comparison be based on? Now here was a question that could be jolly fun to play with.

About this time our veterinary friend, Harry Miller, came by for a visit. I told him, "Harry, we've seen these two male warblers in an amazing battle, and I'm trying to figure out what would be a comparable feat of endurance for a human. Do you have any ideas?"

Harry had doctored animals of all kinds and sizes from a sick tiger and a boa constrictor to a gerbil and even parrots and canaries. But warblers? No.

He thought for a moment. "I have these tables that give us the amount of medication that would be proper for an animal according to its size," he said. "Would they be any help?"

"Do they include humans?"

"Nope." He paused again. "How about comparing the energy involved?" he suggested.

Human and bird energy comparisons? The basal metabolism of a bird changes with the season of the year, with the time of day, and with the temperature.

Joel Carl Welty, in *The Life of Birds,* wrote: "Basal metabolism varies roughly in proportion to the surface area of an animal; hence a small bird metabolizes at a higher rate than a larger one. For example, the basal metabolism of a 27-g House Sparrow is 312 kcal/kg/hr, and that of a young adult weighing 75 kg is about 1600 kilocalories per day, or about 1 kcal/kg/hr."

After reading this statement of Welty's the second time, I thought it seemed feasible that the story of this record battle should be written by a committee of experts.

But while pursuing the energy idea, I had happened on all manner of tidbits too intriguing to set aside. For instance, a British journal states: "Hummingbirds have the highest metabolic rates of any animals — roughly a dozen times that of a pigeon and a hundred times that of an elephant." How often do we get to compare a hummingbird to an elephant?

And there was this in the same journal: "Flying is faster and energetically more efficient than walking or running for comparably heavy animals." To use this method of comparison, we would need a Goldencheek the size of a human or a human the size of a warbler. The latter would make a delightful toy, but alas, Alice's magic mushroom was lost in antiquity.

"Remember, the bird is acting by instinct," Harry had proffered.

In fact, shouldn't the pursuit of this fantasy for entertainment's sake be abandoned, given that the warbler was acting solely by instinct and his behavior was decided at his conception? This is what he *is,* and he was merely doing what instinct ordained. So was it even worthy of a story?

Instinct. An often-used word. Welty also stated: "No animal acts in a vacuum." He went on to say that the word "instinct" "has been freighted with a great variety of meanings, and this makes its modern use at times precarious."

Indeed, some scientists think the term should be abandoned. Nevertheless, the warbler's job is to be the daddy. His goal was in fact noble, for it was nothing less than to give his valued genes for the sake of continuing his species.

Whoa! Let's leave ethics out of this if sex is involved.

"Not so fast," said a member of Planned Parenthood, who just that moment had volunteered to serve on our committee. "Don't you see that this battle between the male birds is a prime lesson from Nature herself? The defeated warbler doesn't have a territory suitable for successfully raising a family, and no female will choose him because of this, and he won't become a parent. The successful male does have a territory that will provide his offspring with good nutrition and other necessities and therefore he will easily have a female choose him. Their future offspring have been provided and *planned* for by Nature's policies for the species."

Well, ah—yes, indeed.

Enough of all this diversion. Let's do this a simple way.

The estimated circumference of the area guarded by Warbler Number One was about 400 feet. For easy mathematics I multiplied that by 80, the number of circles flown, to calculate a total flight distance of 32,000 feet. This is a good starting point. Can we use it to arrive at a comparable situation for a human? Can we use distance to arrive at a comparison?

Well, let's go on. A mile is 5,280 feet, so divide that into 32,000 feet and you get 6.06 miles. Since this is all for entertainment, let's make the math easy by moving it to an even 6 miles. Not an easy zip for our teeny-tiny, itty-bitty bird.

Convert 32,000 feet to inches and you get 384,000 inches. Divide that by the 5-inch length of the warbler, and the answer is that the tiny bird flew 76,800 times the length of his body.

So now that we have a 5-inch length for the warbler, let's choose a 6-foot height for a man. A 5-inch bird and a 6-foot man. Are we getting somewhere? Are we there yet?

Now let's get our 6-foot-tall man in a physical battle with another 6-foot man. Too bad the gladiator days are over. But that sport, like boxing and wrestling, would not serve us, because none of these sports involves distance. So what about a marathon?

Oh, goodie! Now we can get into history. Remember that guy Pheidippides? He's the one who started the marathon thing in 490 B.C. He ran to Athens from a battlefield near Marathon, to tell everybody the great news that the Greeks had defeated the Persians. The victory at Marathon is where we got the name and the race from. Too bad Pheidippides' trainer didn't get him into a proper cooldown—Pheidippides died.

When we say "marathon," we're apt to put "Boston" in front of it. As with so many things in athletics nowadays, we owe this association to a sponsor, namely the Boston Athletic Association, which sponsored the first Boston Marathon in 1897.

As of the year 2004, the official distance for a marathon is 26.2 miles. Drat! Why did they have to tack on 0.2 mile? Well, in our little game we can do whatever we want to, so let's drop the 0.2 and make it easy math.

We already have the warbler paced at 6 miles, and we have our 6-foot man at 26 miles. What we need is how many times each of them distanced his own measure into his traveled distance.

Twenty-six miles is 137,280 feet. Divide that by the man's 6-foot height, and we learn that our 6-foot man has run 22,880 times his own measurement. Not bad.

But compare that to the warbler's accomplishment: this little bit of fluff has traveled a distance 76,800 times the length of his body.

There you are. This brave little endangered warbler has outdone the best marathon man by more than three to one.

OK, OK. These are this story's facts. Don't get all huffed up about it. If you don't like for a man to be physically inferior to a teeny-tiny bird, go ahead—write your own story.

As for me, I say, "Way to go, bird! I knew all along you are a marvel."

HOW BIRD PEOPLE LEARN TO LOVE

"He's a member of the family, and he's in trouble."

In our family this means we will help any way we can, but first we must find him.

Picture an eighty-year-old man with a bum leg and arthritis and his skinny-frail wife, who is not a growing girl, loading up their sixteen-year-old motor home and heading south, a l-o-n-g way south.

Record rains have washed out bridges, turned stretches of highway into unrecognizable slush or to pavement filled with endless potholes, one so deep the couple "almost needed a passport to get out of it."

Now, after various mishaps, the heavily loaded vehicle approaches a final challenge, a corkscrew climb of five thousand feet in only fifty-two miles. In clouds and fog it successfully avoids the descending trucks, and, somewhat remarkably, its two-thousand-mile journey ends at the ancient city of San Cristóbal de las Casas.

In the cloud-touching mountains of the state of Chiapas, Mexico, the old-timers now begin their search for their endangered family member in a forest.

The searchers are Marjorie and Red Adams, and they've had more than a plenitude of explaining to do to both family and friends. People can easily understand and empathize with a strong love and feelings of responsibility for a devoted family dog, but to have the same deep love and feeling of responsibility for a creature that is wild, free, and so different from us that it doesn't know—much less care that we exist—well, simply put, that's *weird*.

"You're too old," our son lectured us.

"That bird doesn't even know you're alive," our daughter pleaded, "so why would you risk your necks for it?"

The answer to that question began more than forty years ago. As I held my binoculars to view an elusive vireo, I was startled to see instead a design in black, white, and gold as elegant as an Oriental painting. The Golden-cheeked Warbler had found me!

These many years later that incident seems fated, for how could I guess that this bit of fluff about five inches long and weighing 0.36 ounce would change my husband's and my life, and how could we foresee that today it would be an endangered species?

"Listen!" Red alerted us suddenly. Our family was sitting at the picnic table deep in the rugged canyon of Roy Creek. Red pointed almost directly above us, and there singing its little *bzzz wee-ah-WEE-zy* of a song was the Golden-cheeked Warbler, announcing that the surrounding real estate belonged to him and no one else.

This rare bird, which had been the revered emblem for the Travis Audubon Society in Austin for many decades, was actually nesting here, with us, on the Pedernales River. We overflowed with a proprietary pride and delight, for not everyone gets this kind of dinner music.

The Goldencheek nests nowhere in the world but about twenty-four counties in Central Texas, so to score it for your North American birding list, you gotta come here.

As the game and sport of birding skyrocketed in popularity, we found ourselves hosts to listers as varied as plumbers, museum directors, genteel elderly ladies, and birding champs and birding bums from around the country and as far away as Canada.

Now the warbler seemed an unexpected gift. With so many people apparently interested in this beautiful bird, we thought: "Why not make a movie about it too?"

Neither Red nor I had had any formal training in moviemaking, so we had fallen by what can only be called an accident into making a wildlife film.

With a series of mischievous squirrel antics and only one line of narration, *Where Should a Squirrel Live?* taught the scientific principle of adaptation, beginning as early as kindergarten level. The movie was successful in three nations and was even aired on Japanese television.

With that success, we amateurs now contemplated making a movie about a bird. As the beautiful little rarity sang above us, we had no idea of what we could be getting into, and we certainly had no hint that this decision would change the compass of our lives forever.

Besides our inexperience and naïveté, the difficulties were numerous. This bird nests in craggy, rough terrain and only in mature juniper-oak woodlands with about a 50 percent canopy. It especially prefers areas where clear springs have carved canyons into the limestone hills. To build its nest, this habitat specialist must also have bark strips from the endemic Ashe juniper tree (locally called cedar), and it wants special oaks for feeding. It is an easy prey to cowbird parasitism and is also victim to such things as snakes, owls, jays, ants,

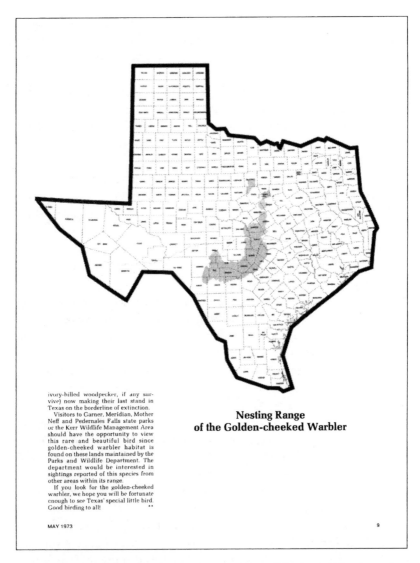

**Nesting Range
of the Golden-cheeked Warbler**

*The Golden-cheeked Warbler's breeding territory is so small that it would be microscopic
on a world map. (Courtesy of* Texas Parks and Wildlife *magazine)*

Every animal has its own place in the world— a place where it lives best.

Where Should A Squirrel Live?

JOHN MUIR AWARD FINALIST

Life Sciences And Ecology

$\mathcal{A}$dams & $\mathcal{A}$dams
FILMS

"Rocky," the playful and mischievous film star, taught the scientific principle of adaptation, beginning at kindergarten level. (Artwork by Zilla and Marjorie Adams)

Red and Marjorie posed for a newspaper photo to celebrate the release of their award-winning film Where Should a Squirrel Live? *by a Hollywood distributor. (Courtesy of Austin American-Statesman)*

feral cats, and the dreaded *development*. Worst of all, it is so active that Red soon nicknamed it Flutterbutt.

Looking back from age ninety-three, when he now weighs in at 135 pounds, Red says, "I guess I had to have a screw or two loose in my brain to strike out at daylight, carrying a ladder through the cedar brakes to climb maybe a dozen trees in a morning.

"The nests are always well hidden, but the problem was not just finding 'em but finding one that could be filmed. I didn't have fancy equipment—a big backpack on my shoulders some of the time—but always that old-fashioned wood tripod and heavy 16 mm camera with its big, extra-long lens.

"Carrying a load on my shoulder that sticks out both front and back and maneuvering it through the brush and up those rocky hills—it was just my fool's luck I didn't fall and break a bone or two. I think the rattlers and copperheads were scared of me."

I, too, searched for nests and locations and spent many hours in research and writing. In all, our labors continued through six Goldencheek seasons.

The result was simple: our film was turned down not only by our West Coast distributor but by all the other distributors too.

We were told with sympathy, "Take your movie back to Texas and finish it the best you can for local use. Maybe you can at least get back the money you spent on film."

Clarence Cottam, director of the Rob and Bessie Welder Wildlife Refuge in Sinton, gave us a $2,000 grant. The Texas Ornithological Society came up with $250. The rest came out of our hides.

The end of this fairy tale, like all the really good ones, was perfect: *What Good Is a Warbler?* was named Conservation Film of the Year by the North American Wildlife and Natural Resources Conference and also by the Outdoor Writers Association of America. It began receiving other honors too. It found its way into schools, libraries, nature centers, museums, and government agencies and to bird lovers in the United States and Canada, along with a few in Australia.

A special honor was its use in a church worship service. The subject— values.

Red, outfitted with wildlife photography equipment, poses for publicity shot.
(Courtesy of Austin American-Statesman)

Thursday
June 14 1990
35 cents

Austin American-Statesman

Weather
Partly cloudy. A 20 percent chance of rain high, mid-90s. Low, mid-70s. Southeast wind 15 to 20 mph. Details, A16

A MOVING PICTURE

Austin pair put love for birds into action with film on warbler

Marge and Red Adams have had their love — and concern — on Austin-area birds since 1985.

Oil spill, placed at

Army to cut 13,000 ...

Austin pair puts love for birds into action with warbler film

Continued from A1

> **Most of the problem we see today with the endangered species is that people are ignorant about them.**
> — Marge Adams

On July 14, 1990, Red and Marjorie found themselves featured on the front page of the Austin paper for their efforts to save the endangered Golden-cheeked Warbler.
(Courtesy of Austin American-Statesman/Larry Kolvoord)

The Adams pair posing for their pictures in the newspaper.

Events change our lives, often in an unexpected way. Those Goldencheek years brought a reward that I would exchange for no other, a rich reward experienced at times by others who are deeply immersed in nature study. Those many hours spent pursuing and studying the Goldencheek revealed that inside that tiny energy machine lives a feisty character that will fight with feather-scattering fierceness to gain and keep a home; a loving and tender mate; a devoted parent that will risk its life to defend its young; and an entity that in miniature does many of the things and has many of the same needs we humans do.

I began to appreciate the marvel it is, but there were so many mysteries. With a bill as its only tool, how could the female create such a beautiful three-inch nest? What deposited in its psyche the necessity for bark strips from a special species of tree?

This question was especially puzzling to me, for during a trip to the Grand Canyon I noticed the shaggy bark on a juniper tree. I stripped a piece of it and held it up to Red.

"Look, this peels off in strips just like our junipers in Texas. So why doesn't the Goldencheek nest here too?"

Every Golden-cheeked Warbler's nest found has been skillfully woven with bark strips from the Ashe juniper, a tree essential to the bird's survival. (Photo by Red Adams)

Obviously, our bird's isolation from its close relatives was caused by more than just tree bark. Food? Plants? Climate? Were even the ice ages responsible?

As I thought of this bird's survival through the creeping centuries, and its life in the ebb and flow of glaciers, time expanded exponentially for me, and I began to see the Goldencheek as an heirloom connected to an unfathomable past and an unseeable future.

Goldencheeks, especially the female, are fond of bathing. (Drawing by Nancy McGowan)

Some of the most unsatisfied questions were these: How does a young Goldencheek know how to fly south to a land it has never seen? And where does it spend the winter? Does anybody really know where Goldencheeks are in winter?

In a juniper-oak grove atop a high limestone hill on our neighbor's ranch, Red had finally found the perfect setup. He spied a female as she frequented an ancient juniper, and in a makeshift blind he shot what he felt was his miracle scene as she pulled the vital nesting material from another juniper. It was one of his happiest days. Then he saw a female gathering caterpillar webs from a plateau live oak and filmed that.

We had found nests in walnut, cypress, and oak trees, but this particular female was building her nest in a juniper close to a tree that was sturdy enough to support a crude blind about twenty feet above ground. Though this blind swayed with the breezes, Red had perfect filming opportunities as the mother brooded the four eggs, taking time off only for short feeding periods or to bathe, preen, or drink.

She flew to and from the nest very directly and with great speed, so it was hard to follow her. The devoted male often brought her food. When the babies hatched, their feeding was shared by both birds. They particularly preyed on large green caterpillars from the Texas (Spanish) oaks and would beat them against a limb to tenderize them before giving them to the young.

In the rain the mother spread her wings to protect her naked babies, and when it got hot, Red saw her remove some of the feather lining from the nest. Once when I came along, the father bird fluttered a distance from the nest and began singing to attract the dangerous human to himself.

Often I stayed near to alert Red when the parents were approaching or to pass things up to the blind. He whispered loudly, "One of the babies is on the edge of the nest. It's going to fledge!"

Quickly, I helped him down with the camera, and we both crouched as he adjusted it low toward the baby, only a few feet from us. Since it had emerged from an egg as large as a bean, it had eaten enough to increase its weight about six times, actually weighing more than its parents. The gawky thing had just enough feathers to help it jump-flit to a low bush.

Despite the camera's whir, the mother filled its gaping mouth with a huge caterpillar. Even here she removed the fecal sac away from the area.

Another baby bounced to earth and was fed almost immediately by its father. As I beheld the brilliant gold on the male's cheek, glowing in the sunlight, he was dazzling.

Soon all the babies were on the ground in the wild world, where they now were prospective food for many foes. Red and I were also a threat.

I could grab it, I realized as one baby moved closer. I held my breath as its mother risked her life to come almost under my chin to give it the necessary nourishment.

"Would I risk my life this way to care for my own babies?" I couldn't help asking myself.

Instantly the thought stripped our species differences down to the elementals. The struggle for a home, a loving mate, successful parenthood. We could match action to action.

Sometimes in the world of nature a sudden enlightenment can come to us that feels like a small connection with the Infinite. A wave of emotion swept over me like a blessing. A door was opening for me to a new world, their world, where my vision had a different glory to it. Time had built a bridge between me and this mother so solid I felt I could walk it. As different as we were, we were moving through life in the same eternal rhythms.

"You don't know it, but I do," I thought. My lips silently formed the words to her: "We're kin."

I was feeling right down to my bone marrow the universal truth that *everything in nature is connected in some way to everything else*. In all my wildlife observations in four countries I have never experienced an emotion as deep as this. These tiny energy machines, making their destined way in the wild world, had been promoted from neighbors to members of our family.

Aldo Leopold said, "Our ability to perceive quality in nature . . . expands through successive stages . . . to values as yet uncaptured by language."

I couldn't explain it, but I can never feel more love for any wild creature than I did at that moment. It is a love that has endured to this day, a love I have puzzled about many times as I compared it with Jane Goodall's devotion to her chimpanzees or David Mech's long-enduring involvement with wolves. Love is love, but the difference between theirs and mine is that the objects of their love finally returned trust, respect, and, in Goodall's case, even affection. The only exchange between me and my Goldencheeks has been an alarm note or flight. The only reciprocation I've received from my Goldencheeks has been an alarm note.

Is this one of the most unselfish loves we humans can experience?

GET OFF MY LAND!

The knock-down, drag-out war began December 27, 1990, and quickly spread to involve all the Texas Hill Country, an area as large as Vermont and New Hampshire combined and with a population of approximately 3.5 million people, including the cities of Austin and San Antonio. Seven other endangered species were involved, but the centerpiece of this great battle was and remains that tiny bit of life previously unheard of by most of the combatants—the Golden-cheeked Warbler. It was declared endangered on that date by the federal government. The battle that this act started still simmers and boils over here and there.

Unfortunately for this small bird, much of the capital city of Austin is smack-dab in the middle of prime warbler habitat, and the city's metropolitan area grew by an extraordinary 45 percent in the 1980s and hasn't stopped yet.

The question seemed to be this: Given that the bird was here long before humans, does it have certain squatters' rights? Commercial interests and many landowners vociferously answered no.

Since every living thing needs a place in the natural world to survive, and since the Endangered Species Act states that destroying the habitat upon which an endangered species depends for survival is a "taking" of the life of that species, thousands of acres of land were legally locked up. For a time the Golden-cheeked Warbler was the most hated and the most loved bird in the Lone Star State.

A furious army made up of developers, ranchers, farmers, highway builders, realtors, homeowners, bird lovers, business owners, and environmentalists swarmed to explosive public hearings. The situation built until dissidents led by states-rights supporters and defenders of the "God-given" rights of Texas landowners to manage their land their own way joined to march on the state capitol. The crowd was at least three thousand strong.

Highlights of the parade were carcasses of cedar trees drug behind trucks labeled with the sign, "The only good cedar is a dead cedar," and marchers carrying placards featuring such messages as a skewered Goldencheek and the note "Tastes just like chicken."

It was rumored that Sam Hamilton, supervisor of the Environmental Services Division of the U.S. Fish and Wildlife Service, had a bodyguard, and it is a fact that he did receive threats.

More than 95 percent of the land in Texas is privately owned, but wildlife in general is owned by the nation. Therefore, in 1991 a large committee headed by former Austin mayor Bruce Todd began a pioneering effort of what seemed an impossible task: saving for posterity the eight endangered species living on the land and in the many caves of this karst-rich and unique biosphere in Central Texas.

Ironically, the birth of the Balcones Canyonlands Preserve would be presided over by realtors, landowners, developers, ranchers, water protectors, environmentalists, taxpayers, industrialists, citizen watchdogs, and bird lovers who were all fiercely fighting separately to protect hotly antagonistic interests. It was only the powerful Endangered Species Act, which eventually required them to set aside at least 30,428 acres of habitat as a permanent preserve, that forced them to cooperate.

The price for large landowners can be illustrated by the situation of the 3M corporation. A major concern in the Texas Hill Country is water purity and supply, and before the first spade of dirt was turned on 3M's proposed Austin Center on the outskirts of Austin, the company undertook studies with the City of Austin that focused on preserving water quality. Then, when the Goldencheek was declared endangered, an estimated eight nesting territories of the warbler were found on the 3M building sites.

"Our corporation has expended in excess of a million dollars to donate a two-hundred-sixteen-acre tract of land as mitigation for the eleven acres of warbler habitat lost," Bill McClellan, 3M staff vice president, told me in 1992. "In addition 3M has spent fifty thousand dollars to revegetate the new tract, and we created an endowment of fifty thousand for management of the new preserve. We also funded a three-year warbler census and Brown-headed Cowbird trapping program. We estimate that at the end of ten years the cost will total almost two million dollars."

Another costly mitigation involved the FM Properties Operating Company. Through a series of purchases and exchanges in cooperation with the Nature Conservancy of Texas, the company contributed more than four thousand acres of land, absorbed slightly more than $8 million of a county road construction debt of approximately $28 million, and paid for at least twenty years of reserve management: a total of approximately $20 million dollars.

The Schlumberger Austin Systems Center even went so far as funding nature trails on its four-hundred-acre campus and an alarm system that alerted for feral cats so that they could be removed.

Separate from this hard-won preserve, the U.S. Fish and Wildlife Service began the Balcones Canyonlands National Wildlife Refuge with the purchase February 25, 1992, of 630 acres, which is slated to grow to at least 46,000 acres. By September the same year Dean Keddy-Hector of the Texas Parks and Wildlife Department and Carol Beardmore of the U.S. Fish and Wildlife Service prepared a recovery plan for the Goldencheek.

These deeds would potentially assist the Golden-cheeked Warbler in surviving on its remaining breeding grounds, but saving half a bird's life isn't enough. What happens to the species in winter?

The Goldencheek is only one of more than 350 species of our cherished North American birds that spend their winters in lands or waters to the south. As a group they number into the billions. A thirty-year study shows many species are declining, some at an alarming and increasing speed.

To meet this crisis, the nonprofit National Fish and Wildlife Foundation began organizing in 1990 a new cooperative called Partners in Flight (Aves de las Americas) to expand the study and protection of migrant birds to include their winter homes in other countries.

Now Marjorie and Red Adams asked for and received help from 3M, and the National Fish and Wildlife Foundation matched it. If these birders could find the Goldencheek in Mexico, could they learn ways to help it survive? If so, would it be possible to add the other half of the bird's life to their film so its story to the world would be complete? Such were their thoughts as the old couple began their exploratory journey south in 1992.

Since the first Golden-cheeked Warbler was collected in Guatemala on November 4, 1859, by British ornithologist Osbert Salvin, few winter sightings had been recorded anywhere and there was strong dispute about the bird's winter habits. Even its nesting grounds were not known until some years after the first U.S. record in Bexar County, Texas, in 1863.

Now the news was that Russell Greenberg, director of the Smithsonian Institution's Migratory Bird Center, and his team of bird specialists from the Smithsonian had reported the Goldencheek for two successive winters in far southern Mexico in the state of Chiapas.

When Red and I arrived there, John Sterling and Peter Bichier Garrido were censusing all the birds of the area. John had chased birds for himself and

The Adams pair ready to take off for another adventure in their motor home.
(Courtesy of Lew Adams)

for organizations in ten countries. Peter, born in Venezuela of French parents and schooled in Colorado, sometimes got so roused by tough environmental problems that he lapsed into a three-language accent.

In answer to our questions, John obligingly pulled the records from his computer and made this announcement: "Out of one hundred observation points, we saw or heard Goldencheeks only four times. That's four percent."

Red's glum East Texas translation was "No, it's looking for four fleas on a hundred dogs."

Then exactly how do you search? In Mexico the word for "warbler" depends on where you are. In Spanish the Golden-cheeked Warbler is called Gorjeador cariamarillo or Chipe Mejilla Dorada, but most bird people use the word *chipe,* meaning, in general, "small bird." None of the seven Indian languages in Chiapas includes a word for "small bird."

So we began our conversation with Flavio Martinez and Petra, his wife of fifty-seven years, by pointing to the picture of the Goldencheek on our postcard and asking, "Have you seen this bird?"

The family courteously took the matter seriously, and Flavio passed the card to his son, Xavier, who studied it closely.

"I have seen many small birds. Some with yellow. Perhaps I have seen it," he said encouragingly.

Beyond the servants' quarters, where we sat, blazed the white walls and turquoise blue shutters of a four-hundred-year-old villa. Flavio looked after it and the surrounding acreage for the absent owner.

"When I came here young, nearly every tree had been cut down," the old man remembered. "All was used for grazing, and much land was washing away. *El señor* wanted trees, and I planted at least two thousand. This forest was made by *me*."

The trees in his forest had now grown tall, and the path through it had been trod by so many animal and human feet that in places it was almost three feet deep. Next week we would get a brief glimpse of our first Mexican Goldencheek there, but for now we were seeking information and asking the ordinary get-acquainted questions: What about your family? Flavio and Petra had five sons and two daughters. And as one grandparent to another, How many grandchildren?

Flavio fell silent; then floundering, he deferred to Petra. She counted on her fingers, tugged on a gray braid prettily twined with pink ribbon, and pondered. He mentioned another son's children, and their count continued. Finally, they answered, "Somewhere between thirty and forty."

This was our introduction to the population explosion in Mexico and other countries south of the United States. Food must be found for each new mouth.

Miguel Alvarez del Toro, director of the Instituto de Historia Natural (Natural History Institute) and a lifelong protector of wildlife and the environment, had given our project his full support. On a cold and foggy morning Eduardo Morales Pérez, chief of the Department of Conservation at the institute, drove us on a long search.

Much of the red earth in some sections was completely bare, but there were signs of reforestation. However, many of the small trees had been yanked out of the ground.

"The people want crops, not trees," Eduardo explained. "They refuse to understand that this soil can't grow crops."

In due time we reached a hillside where we were surrounded by a full-grown forest, which was lavished with bromeliads, ferns, and blossoms. I stared at its complex beauty, feeling privileged, and in an ethereal light it seemed we three stood within a painting, participants in a masterpiece.

We walked a gentle climb in this ancient and diverse forest, but almost at once we came to an opening where it seemed a gardener had gone mad and scattered his work everywhere. A tree theft had left bromeliads, ferns, orchids, and other lush plants without a home.

As we climbed higher, we found a stump more than six feet across and not the least decayed, so this ancient giant had been alive in recent times. We had not seen a living tree this large in all our explorations.

A Mountain Trogon called in the distance, and almost immediately a rooster crowed nearby. Its domestic call seemed an appropriate warning, for without guards the remainder of this magnificent forest will almost certainly disappear.

On another morning, cold and foggy, Eduardo took us out the four-lane Pan-American Highway to a road that quickly became a rocky ditch, and then we were confronted with what Red estimated to be a 20 percent climb.

Pressing the gas pedal, Eduardo tackled it, but the cobblestoned hill was glazed with frost and the truck slid back toward a twenty-foot drop-off. On our second try the tires successfully ground us to the top, and we traveled a long way on a nearly impassable road leading to an Indian village.

At Mount Peña Maria, where he had seen a Goldencheek the year before, Eduardo stopped, then commented with little surprise, "Every tree is gone." We could hear an ax chopping in the distance.

We had just gotten good studies of two Rufous-collared Sparrows when hooves sounded on the rocky path. We moved aside quickly to let pass a small burro hidden by its skillfully packed load of wood, each piece about the same size and length. Beside it a family of Indians in traditional costume trod along in the floating pace of the heavily laden, their huge load of wood or other goods tumplined from their foreheads to rest on their stooped backs. Even the smallest child had her burden. They exchanged greetings but didn't turn their heads or lose tempo.

It was clear. There goes the forest.

Eduardo read my thoughts and said gently, "They are people and they must eat."

We saw slingshots in all the markets. The Natural History Institute exhorts youngsters not to slay birds and small animals, yet a bird or even a mouse may be added to a boiling pot of beans.

Mexico has the largest population of native peoples in the world, and its people are as diverse as the lands they live on. Through the centuries there have been countless wars between tribes and cultures, with the vanquished often enslaved. Under the Spanish, slavery became almost universal, and in some isolated areas change has been glacial.

The Zinacantecs are typical of the Indians' problems. In 1960 they numbered 7,500. By 1992 they had grown to 25,000 strong but were squeezed on the same amount of land. It is owned by the tribe, with no individuals holding titles. Fertile land in the lowlands, where the men used to raise the additional food needed to feed a family, has been taken away by cattle ranchers, often through force.

What happens to the land depends on what happens to the people, so to learn more I visited several times with Robert Laughlin, curator of Meso-american Ethnology at the Smithsonian Institution. He divided his time be-tween the United States and his home here in San Cristóbal.

"When I came here thirty years ago," he told me, "Indians were scandal-ously abused and even robbed with impunity. Any Indian caught in town after dark was jailed and fined, and they were not even permitted to walk on the sidewalks. Now, with the historical city a prime tourist attraction opened to the world, the Indian situation is much improved."

But many of the native peoples had no steady income and had to depend on crops or the sale of firewood or charcoal.

I asked him, "What are the Indians' values concerning the environment?"

"The Maya have a deeply held belief that they are a part of nature," he said. "This belief permeates their lives and has endured through all the many ideologies that have been forced upon them. They consider themselves true Christians, but their pre-Columbian gods and ceremonies are present in their everyday surroundings and thus are the most influential. They both revere and fear the natural world, believing that human survival depends on harmo-nious relations with the deities. They believe their work on the sacred land is their repayment to God for giving them the land."

I learned that their most powerful deity is the Earth Lord, for he is present in every natural element from wild beasts to the caves of the underworld. He controls all of nature and fights to keep the land wild. When the land is used by people, the Earth Lord must be propitiated, and there are many ceremonies used to satisfy this powerful deity.

One ancient belief is that every tree has a spirit. The trees call the rains; the rains create springs and water for all creatures. The trees give warmth and fire for cooking, and trees give shelter and shade.

A Mayan admonition given to me by Brian Stross, an anthropology pro-fessor at the University of Texas, goes something like this:

> The old father said to his son, "Always listen
> to the song of the Toucan:
> *Te'etik pahal sok te kuxlehale*
> *Tame ya x yal spisil te'etik ta lum*
> *Ya me x yal a'kuxlehal ek.*
> Trees are life.

If all the trees fall,
So will your life."

Can a hungry man afford to listen to the song of the toucan?

The good news is that several organizations are now involved in reforestation, and in this climate even fence posts sprout.

In 1991, John Terborgh—scientist, professor, and author of the book *Where Have All the Birds Gone?*—stopped over in San Cristóbal after a trip to the Montes Azules Biosphere station in the Lacandon jungle, and we had a good visit. He had firsthand knowledge of the problems of the unemployed poor from his visits in fifteen countries of the Western Hemisphere.

"The poor have no choice but to find a patch of unoccupied land where they can raise enough food to survive," he said. "No place is too remote or too forbidding to a man who is starving, and he will clear and plant even land not suitable for agriculture. If NAFTA allows free trade nationwide, it will mean new job opportunities, giving impetus for people to live in the cities. I believe we still have forests and wildlife in the U.S. and Canada because the majority of our people live in cities."

"Not so," Duncan Earle, anthropology professor at Texas A&M University, told me in the late 1990s. "NAFTA was the trigger that set off the most deadly revolt in Mexico in eighty years." Earle was accepted by the Chamula Indians in Chiapas and lived with them for several years, learning about indigenous life and problems firsthand.

"NAFTA faced the poor with still another threat, for they don't trust Americans any more than they trust the cattle ranchers who have taken their land. In fact, some say this will be the end of life as the Indians knew it. The revolutionary constitution of 1920 guaranteed land reform by limiting the number of hectares of land any citizen can own, but Chiapas has been the forgotten state, and the revolution is just now arriving there."

With the force of the Zapatista National Liberation Army offering protection, some Indians moved back onto lands that had been theirs for centuries. The angry response from the cattlemen and their private armies was to promptly shoot the Indians.

Peter Ward, Mexican specialist at the University of Texas Institute of Latin American Studies, had this to say: "Actually, in recent years Chiapas has been the second-largest recipient of government assistance, but most of it has gone to *ladino* [non-Indian] communities or into the pockets of local political bosses. Obviously, both political and land reform are needed."

It was Andres Sada—an industrialist, Mexico's champion birder, and author of a book that keys Mexican birds by the difficulty in finding them—who introduced us to Pronatura. This is Mexico's first private, general-membership organization dedicated to the preservation of biological diversity and the protection of endangered flora and fauna.

Rosa María Vidal, director of Pronatura Chiapas, took us on our first search for the Goldencheek in its wintering grounds.

Red and I kept our binoculars at the ready as Rosa led us down a narrow path one morning. "This time I'm not dreaming," I thought, but this valley so far from home matched none of my dreams. We were in the tropics, but at six thousand feet elevation there was thin ice on the shrubs. These old fields with clumps of second-growth trees and these steep mountains covered with pines and oaks on each side were not what I had expected. Of course, I reminded myself, we knew it wouldn't be like the Texas Hill Country, but this place didn't have the *feel* of Golden-cheeked Warbler habitat.

Nevertheless, Rosa was saying, "They've been seen here."

Red sighted a Rufous-collared Robin in a blooming peach tree, but I ignored it. I had a glimpse of a bird with yellow on its head, and I stood as fixed as a post. My decades-old dream could come true this instant!

But it was a Townsend's Warbler, and my heart was slowing its thumping. I had wondered how I would behave if I finally saw my beloved warbler living the other half of its life, but now I knew I wouldn't cry.

Rosa had explained to us that Pronatura's landmark achievement was the creation of a legal method by which private nonprofit organizations could own a protected preserve. One of these is Huitepec, a 135-hectare reserve of fragile and endangered cloud forest still surviving on three mountains, all above seven thousand feet in altitude.

At Huitepec, Claudia Macías Caballero was working on her master's degree under the supervision of Pronatura's Vidal. Macías Caballero's project was the study of different birds' use of honeydew. Just to reach her observation area meant making a climb in steep and rugged habitat in the high altitude. Here she also had seen the Goldencheek. Red and I were not lucky enough to find it there.

Pronatura and other organizations and individuals striving to save what they can of natural habitat and wildlife in Chiapas face a formidable task, partly because of language, educational, and cultural differences and partly because 60 percent of the population lives in rural areas at, or barely above, subsistence levels. We saw several shantytowns dug into unstable rocky hillsides on land the occupants didn't own, and all without water or sewer. One report stated that fifteen thousand people died of starvation in Chiapas in 1992.

In comparison with the people's poverty, the land is astoundingly rich.

The Nature Conservancy states, "Chiapas is one-third the size of California, yet it contains 641 species of birds, almost as many as occur in the entire U.S.; over 1,200 species of butterflies, and 17 natural communities extending from mangrove swamps to cloud forests containing 8,000 known species of vascular plants . . . a total number of plant and animal species unmatched anywhere on the North American continent."

If the forests can be saved, so also will much of this biological heritage. We were told a law had been passed to prohibit tree harvesting and the jails were full of violators. This was followed by the statement, "But the ones doing the most cutting are free."

We met Paul Wenninger, a young American biologist stationed at the Montes Azules Biosphere where he was studying the use of trees by different birds. He told us that three of the biosphere's scientists had identified the songs of three hundred bird species there, but they had heard no Goldencheeks. Revolution had gone on in Guatemala for more than three decades; thus in the jungle here, so near the border, the researchers could hear bombs every day and they never dared to wear green or anything resembling military uniforms.

"Come look!" our American neighbor called to us one afternoon. We heard a strange throbbing cadence and rushed the few yards to the highway. The wide pavement was covered from side to side and from horizon to horizon with solemn yet somehow festive men, marching with the purpose and steadiness of army ants. Trucks and cars accompanying them were so encrusted with humanity that they seemed sure to collapse, yet another one or two men somehow managed to squeeze aboard.

A few days later the Pan-American Highway was blocked by an angry mob, demanding land and the end of human rights abuses. As traffic piled up and threats became roars amidst the honking horns of tourists, an inspired officer at the adjacent military base broadcast a gracious invitation to all to join him for dinner. As a meal was spread, the confrontation was reduced to something closer to a fiesta.

Luckily, this was not the day Red and I searched for the Goldencheek in the vicinity. We didn't know it, but this protest march was only one of many preludes to the fateful New Year's Day 1994.

We had searched for two months and had seen many of our cherished North American birds wintering here. John Sterling and Ernesto Ruelas Inzunza had seen a male and a female Goldencheek, and Claudia Macías Caballero had seen a female in her study area. She had also seen a male in migration at her house, only a few blocks from downtown San Cristóbal. Red and I had sighted only two Goldencheeks, both at a disappointing distance.

Everywhere we had gone on this quest the Mexican people, total strangers to us, had been helpful, courteous, and generous. With the ancient citadel blooming pink, white, and yellow in every direction, we couldn't help but feel a certain loss as we joined the migrating birds to begin our own migration north.

"How can we help Goldencheeks if nobody can find enough of them to learn what is best for them in winter?" I complained.

"We haven't made any more headway than a Model T stuck on a watermelon rind," Red summed up. "Maybe it's time to dial 911."

Now we could risk our heavy vehicle on the winding mountain road that Dr. Alvarez had discouraged us from attempting in the rainy season. We found Laguna Bélgica to be a jewel of a nature reserve, a 47.5-hectare island of abundant semideciduous tropical jungle and fauna protected twenty-four hours a day by three knowledgeable attendants.

It proved no inconvenience to park the motor home in the entry road, the only level spot this mountainous terrain offered. Vegetation was so thick that we scored a Life bird, a Red-throated Ant-Tanager, only two feet from our window.

Here in rugged and broken terrain with limestone-based soil, oaks, many trees only medium height, a full supply of thickets and understory, and, of course, water, we felt that at last we were in perfect habitat for our Goldencheeks.

Santiago de la Cruz, the chief guard, assured us that the area was always this green, and he showed us a medium-sized tree that Golden-cheeked Warblers were strongly attracted to when it bloomed.

"Oh, yes," he said. "They may come as many as ten or fifteen at a time!"

I stared at Santiago's face, trying to read it. Had we heard this man correctly? So many Goldencheeks at once? And why would Goldencheeks be interested in flowers? For insects? Nectar? Honeydew?

"Yes, this one," Santiago said, as he promptly pointed to our warbler in the guide book.

"Jean Moore was right!" I exclaimed. "Her reports can be trusted."

"Yes, this lady stayed in the cabin for long visits," Santiago said. "She kept records every day, and we helped her to get a list of all the birds for all the year long."

Santiago had an address for her in Davis, Arizona, but who *was* Jean Moore? Her reports of numerous Goldencheeks, handed to us by Andres Sada, had brought us here in the first place.

Back home again, in a few months I would track down Jean Moore in Prescott, Arizona, and learn she was a solid example of the growing scientific usefulness of information gathered by qualified amateur birders around the world. She had been involved with such projects as raptor banding and a study of the nesting success of Double-crested Cormorants. She had lived in Mexico as a child, thus spoke Spanish, and had volunteered to work with the Natural History Institute on several projects. Her most ambitious was a fifty-four-page *Annotated List of the Birds of Laguna Bélgica*, accomplished in studies dating from December 1985 to May 1988. She recorded the Golden-cheeked Warbler from October through May and included this note: "Had I not seen it also at Mirador El Roblar on Cerro Sumidero, I would think that all the birds from its very restricted breeding ground on the Edwards Plateau in Texas had come to the Reserve."

But as we camped in Laguna Bélgica in April 1992, we had not yet seen Jean's full report. In this natural Eden, the special trees had just finished blooming and Goldencheeks were in Texas singing, building nests, and making baby birds, so we took time now to enjoy beauty, new birds, and the stories of Santiago, his son Abelardo, and the third guard, Eneas Morales.

All of them had worked to build the well-designed trails and the occasional bench, and all knew well the birds and haunts of Laguna. This place was heaven.

In such rugged country we octogenarians carried light stools and sat in first one choice spot and then another. On a side road cut like a gash in the mountainside, we gazed down at least a mile at an emerald valley dotted with toy cows. Birds came and went, and some were new to us, and here came a flock from far below.

I gasped, "It could be him, Red. Look *hard*. Don't lose him!"

A flock of small birds came into trees on the edge of the bluff, and he was with them. In his warbler way he was flitting from branch to branch, feeding his daylong appetite, oblivious of us as he always will be. In this paradise setting, our sighting was one to remember forever.

A voice was saying, "You precious miracle of fluff, don't you know you're too little to fly so far from home?"

There I was at last on a mountainside in Mexico, talking baby talk to a member of my family. My heart was racing but couldn't keep up with the rapid beat of a small bird's heart, throbbing with the ages-old call of spring in other lands.

Then my chest tightened as I thought of the dangerous miles ahead. "It's so far, and you're so tiny," I called after him. "You *have* to make it, because we'll be waiting for you."

"I see *her!*" Red exclaimed.

It was a short but joyous family reunion, a fine climax for a decades-long love affair, as Red and I stood gaping at each other, each knowing exactly how the other felt.

"Good heavens!"—it dawned on me—"that could be the same pair we watched at Roy Creek last spring."

Here at Laguna Bélgica, with its steep ascending trails that allowed views directly into the special trees, we had found a perfect place for filming the Goldencheek in its winter home.

But it was not to be.

"Red, they've got Peter!" I called into our dining room at home. "They've arrested Peter!"

The newspaper's headline read, "Mexican Military Mistakes Birdwatcher for Rebel Leader." The story elaborated: "Peter Bichier Garrido said he was an ornithologist. The Mexican government said he was the mysterious Comandante Marcos. . . . Bichier has been threatened by interrogators, who thought he was concealing computerized rebel information."

Was this the return of "Mexico Bronco" (Unruly Mexico)?

I called Russell Greenberg at the Smithsonian. "Peter is all right," he assured me. "But we had four pretty terrible days at our house in Ocosingo."

When the Zapatista National Liberation Army attacked the government in San Cristóbal on New Year's Day 1994, the rebellion quickly spread across the countryside. In the house in Ocosingo where the Smithsonian scientists were headquartered, they could only listen anxiously to gunfire and hope for safety as they saw men falling dead outside their windows. The streets were soon littered with bodies. Apparently a post had been set up around the scientists' house, and they were spared.

"On the fourth day, it calmed enough for Peter to drive the rest of us to the airport in Tuxtla Gutiérrez," Greenberg said, "so we got out of the country safely."

Peter then headed back to Ocosingo but was stopped and arrested. He had green eyes, he could speak several languages, and he fit the description of Comandante Marcos, the spokesman and leader of the rebellion. In the course of interrogation, Peter was handed over to two husky men who informed him they were taking him to a place where no one would be able to hear his screams. Then suddenly he was released.

Back at the house in Ocosingo, Peter had found it almost cleaned out, and all the electronic equipment and camping gear gone. In the backyard he began digging and retrieved the computer he had wrapped in plastic and buried there. All of the team's two years' worth of data was saved. With the help of two Mexican volunteers, Peter worked another month to finish as much of the bird census as he could before spring migration took birds north.

Coppelia Hays, Latin American coordinator for the U.S. Fish and Wildlife Service, told me that two proposals from Chiapas for Golden-cheeked Warbler studies were on hold, and everything in the area was closed down.

With conditions too dangerous for fieldwork, there would be no filming of warblers by Marjorie and Red in Mexico.

THE CLOCK KEEPS TICKING

There were no balloons or fireworks, but it was a gala day October 16, 1992, at a far-from-anywhere place in the Texas Hill Country. The U.S. Fish and Wildlife Service and the Nature Conservancy were hosting a wild hog (javelina?) barbecue. The menu appropriately included cakes with accurate portraits of the Black-capped Vireo and the Golden-cheeked Warbler created by Diana Johnstone, the wife of a Fish and Wildlife Service employee. Red Adams got the vireo's head.

Leather-faced ranchers in fake Stetsons, genteel ladies with diamond earrings, and VIPs in business suits rubbed shoulders with brown and khaki uniforms as white-haired veterans of World War II presented the colors and we repeated our allegiance. My chest was expanding with glorious patriotism as I glanced down to see my feet touching earth that was already part of the new Balcones Canyonlands National Wildlife Refuge. The crooked and aged oaks that shaded us, the shaggy-barked junipers, that limestone bluff the Cliff Swallows stick their nests on, and the long vistas from these baby mountains were ours to be a forever home for a golden-cheeked member of our family and, under this bird's protective umbrella, all manner of other living things for generations to come.

Congressmen Jake Pickle of Texas and George Miller of California had gotten $8 million in the budget for the refuge, and they both made speeches. Miller began his with "This is my first endangered species meeting where there hasn't been a fistfight."

Some of those present didn't join in the laughter. Standing in front of me was the rancher who only twenty minutes ago vowed, as only a Texan can, "I don't want anybody telling me what to do with my land, and I will *never* sell."

"Don't worry," I told myself. "Fish and Wildlife has time and mortality on its side, and the willing seller will eventually come. Listen, mister, we've won!"

We rejoiced to have our own Jake Pickle take such a pivotal role. For years we had prodded him to be pro-warbler, to the point that at one meeting he said, jovially enough, "Marjorie Adams has been a thorn in my side for years." Then he added, "I keep reminding her, warblers don't vote." Now times had changed enough that he could stand proudly on the podium for this accomplishment.

Things were happening in Mexico too. On October 22, 1992, Red and Marjorie attended the Fourth International Symposium for Wildlife held at the Universidad Autónoma de Tamaulipas in Ciudad Victoria, Mexico. Jane Lyons, project director of the National Audubon Society, was giving the first-ever presentation of a scientific paper on the Golden-cheeked Warbler outside the United States.

Our purpose in being there was to support Jane and to film her historic presentation, delivered in Jane's perfect Spanish. Red and I had to get the English translation through our earphones. We then had visits with various attendees, some of whom reported their own sightings of the Goldencheek (of course, no way to check these), and we passed out our Goldencheek postcards and gave all the publicity we could to the gathered scientists and students.

Noteworthy for the trip was the great number of Scissor-tailed Flycatchers along the latter part of the journey south.

In her same capacity with Audubon, Jane acted as an intermediary in January 1993 for the U.S. Fish and Wildlife Service and met with organizations in Mexico, broaching the possibility of establishing a "sister" park or reserve in the state of Chiapas to cooperate with the Balcones Canyonlands National Wildlife Refuge in Texas.

In April 1993 the indefatigable Jane arranged for biologists Francisco Martin Gomez from the Institute of Natural History and Leonardo Corral from Chiapas to come to Texas, and she personally led them in field studies of the Goldencheek on its breeding grounds in Central Texas. Thus the scientists got acquainted with the summer behavior of the warbler they had been studying in the winter in Mexico.

Jane also made a brief survey of Goldencheek habitat in Guatemala for the Texas Parks and Wildlife Department in 1994, but she sighted no birds.

In the winter of 1995, Daniel Thompson, under contract with the U.S. Fish and Wildlife Service, made a wide-ranging search in both Guatemala and Honduras. In little more than a month he and several helpers sighted thirteen individual Goldencheeks. The birds were seen feeding mostly in oaks.

At Lake Catemaco in Veracruz, Red and I had met William Schaldach Jr., the way birders often meet each other—in the field. Mr. Will, as we called him, had kept ornithological records in Mexico for years, and when we asked him if he had ever seen a Golden-cheeked Warbler, his records revealed one observed years ago in northern Mexico in an area south and east of Big Bend National Park.

When we mentioned this to Andres Sada, he said, "Yes, it's possible. I know an area that has junipers and oaks where the warblers could even nest."

He promised me he would go there in the near future. In December 1995, Chuck Sexton, a biologist with the Balcones Canyonlands National Wildlife Refuge, met Andres in Mexico, and they flew over this territory to pick the most likely habitats. Then in April 1996, Chuck and Cliff Ladd, Natural Resources Program manager for Travis County, Texas, met Andres in Mexico, and together they flew to the area south of the U.S. border in the state of Coahuila.

"It's a land of very large ranches," Chuck told me, "and the ranchers' method of getting around from one ranch to the other is to fly. There are many canyons in the Serranias del Burro range, and we found Ashe juniper trees in them accompanied not by the Texas oak but by the Chisos oak. There were also big-toothed maples and even dogwood. It looked like good Goldencheek habitat.

"One rancher would help us in his area and then pass us along to another rancher, so both in the air and on the ground we covered quite a bit of territory. Unfortunately, the land was suffering the severe drought of '96, and very few birds were showing breeding activity.

"We didn't find the hoped-for new colony of Goldencheeks, but the area should be searched again in a more favorable season. However, the combined canyon habitat probably couldn't support more than a hundred pairs of warblers."

In 1996 the results of a study of the Golden-cheeked Warbler in the highlands of northern Chiapas made by Rosa María Vidal and Claudia Macías Caballero of Pronatura and Charles D. Duncan of the University of Maine were published in *The Condor*. The report detailed forty-six sightings of Goldencheeks, of which thirty-three were considered wintering birds. Next to the very rare Pink-headed Warbler, the Goldencheek was the least numerous. The warblers especially used pine and pine-oak forests, and, as in Texas, gleaning was their feeding method. Therefore, after long years of questions, this landmark study finally officially verified at least one winter Goldencheek home in Mexico.

There had been conflict as Balcones Canyonlands National Wildlife Refuge was founded, and the founding of the Balcones Canyonlands Preserve was no different. As expected, the atmosphere of trench warfare in the Austin area meant that compromise between biological and socioeconomic factors was a continual wrestling match moving at a glacial pace. Though the Balcones Canyonlands Preserve began its physical existence January 8, 1993, with the purchase of 5,280 acres, it was not until 1996 that a Section 10-(a) permit was granted by the U.S. Fish and Wildlife Service to officially establish the preserve.

The preserve management was structured with the City of Austin and Travis County as joint permit holders and with the Lower Colorado River Authority, the Nature Conservancy, and Travis Audubon Society as managing partners. As of September 2004, only 2,875 acres were still needed to complete the 30,428-acre preserve.

The major problem with the original structure was that Travis County was not able to pay its expected share of $45 million for planned development, because the county bond issue failed in 1993. Since then the preserve has evolved as a mosaic of public and private mitigation lands rather than an exclusively city-county system.

"However, the situation for the preserve looks much better now," Mel Hinson, environmental program supervisor for the Balcones Canyonlands Preserve, told me in October 1998. "The county has set up a separate tax benefit account to which funds will go whenever property in habitat undergoes development under the plan. With the price of potential preserve land going up, so will the annual proceeds to purchase the remaining land. With this expected revenue, new lands can be purchased as soon as funds are available. Also, revenue bonds could be considered to finalize the preserve system.

"Another change was enacted this summer," he continued. "The original mitigation fee per acre for warbler habitat was set at five thousand dollars. This has now been reduced to three thousand per acre for a one-year period to encourage developers to participate sooner. This fee reduction has recently brought more money into the program for land acquisition."

"Are there still grudges and bitterness among landowners?" I asked Hinson.

"Yes and no," he said. "The landowners in the designated preserve area can live on their land and use it, but they have had a hard time selling it for certain development purposes. It has been nearly eight years since the Golden-cheeked Warbler was listed, and they understandably feel they are held hostage. They turned to the Texas Legislature to force some kind of relief without success. However, with the new funds from the county program to purchase land and with increasing private mitigation arrangements with U.S. Fish and Wildlife, landowners should get relief faster."

Federal funds are expected to be appropriated to increase the size of the

Balcones Canyonlands National Wildlife Refuge by an additional 15 to 20 percent. When all of the land is acquired, the national refuge and the city-county preserve will have set aside a combined total of more than 75,000 acres so that future generations can still know and enjoy the Golden-cheeked Warbler and seven other endangered species, as well as dozens of other species of concern.

Both private and public preserve programs are required to report annually to the U.S. Fish and Wildlife Service, which issued the permits. Monitoring of the bird populations has begun so that viable populations can be sustained in perpetuity. Even so, habitat can be unobtrusively changed bit by bit by "natural" causes, and constant delay can dull determination and will.

There is no doubt that for some landowners who have been unable to develop or sell their land because of restrictions enforced by the Endangered Species Act, it has been a hardship.

One such owner, Robert Brandes, told me, "It was not the money so much as the inequity that made me fight the rulings of the Endangered Species Act. The Balcones Canyonlands Conservation Plan [BCCP] committee was loaded with developers and environmentalists, and I was the token landowner. For about twenty years I had owned my 150 acres on Lake Travis, miles away from the planned Balcones Canyonlands Preserve, but my land was Goldencheek habitat. That meant it was subject to the $5,000 per acre mitigation fee. This immediately put a financial burden on any buyer but also a limit on the percentage of the property that could be developed.

"Fairness was not a part of the equation for the BCCP, and I felt a concern not only for myself but for all the landowners in western Travis County. I think the Fish and Wildlife Service was suffering from environmental greed. It was obvious from the beginning that the glowing projection that the total cost for the project would be thirty million dollars was unrealistic. One example, and there were many, was that there was no mention of the cost of operations and maintenance.

"As it happened, I talked on national TV with Bruce Babbitt [then secretary of the interior] when he visited the preserve. He told me the right thing for the government to do was to buy out the landowners. Of course, that never took place.

"And after the BCCP committee had met and wrangled for many years about the preserve, I finally told Bruce Todd, the chairman, that I would stop fighting the plan if the objections and concerns of the dissenters like me were included in the final plan report. As it turned out, our side of the story was never included in the report."

Brandes handed twelve boxes of records concerning his battle and that of others to the Center for American History at the University of Texas.

"The Endangered Species Act destroyed my plans and investment for my retirement and old age," Margaret Rector of Austin told me.

I remember Margaret from the days of the Great Depression when she came to Austin to work in the state capitol. We were all poor in those days, and we learned firsthand we'd better provide for the future. Margaret's plan was land, and with monthly payments she bought fifteen wild and wooded acres west of Austin. As the city grew, the land, most of it on a busy highway, leaped in value to a possible $900,000.

Margaret had a buyer—and then the Endangered Species Act took effect. Her land was designated probable habitat for both the endangered Golden-cheeked Warbler and the endangered Black-capped Vireo. Suddenly her parcel would require $5,000 per acre in mitigation and only 41 percent of it could be developed. Her buyer defaulted, and the land was offered for sale on the courthouse steps. There were no bids, so ownership of the land returned to her.

"Amazingly, I suddenly became the 'poster girl' for people and organizations who were fighting the Endangered Species Act," Margaret told me. "Fox Television began to support me, and I told my story on *Good Morning America*. I was the poor hardworking office clerk who was cheated of financial security in my old age by a mean and unfair law, and my story was told in newspapers and national magazines like *Time* and *Nation's Business*.

"I wrangled and I fought to get my land free, and finally, tired of lawyers' fees and exorbitant taxes, I just gave up and practically gave it away."

John and Kim Vaught had treasured and protected the natural beauty and wildlife on their one hundred acres of the Hill Country for 40 years. In fact, in their zeal to save wildlife they once offered a portion of their property to the Nature Conservancy. But Kim said, "The irony is that our land—along with our future—has been taken away from us." All because of the Golden-cheeked Warbler and the Endangered Species Act.

The Vaught property isn't in the actual Balcones Canyonlands Preserve acquisition area, but it is next door and has been designated as confirmed warbler habitat. The Vaughts say this designation has surrounded their property with so many restrictions that it has rendered the land unsellable.

After a developer offered the Vaughts more than $1 million for eighty-three of their hundred acres, they learned there was "a probability that only a small portion of the eighty-three acres could be developed in exchange for the remaining property to become part of the preserve." Developers then offered to pay the environmental mitigation fee for the Vaught acres or even to buy an off-site property in exchange for it, but the Vaughts said both offers were flatly refused.

Then city and county representatives "advised" the Vaughts to pursue a 10(a) permit from the U.S. Fish and Wildlife Service to avoid paying the "prohibitive mitigation fees" required by the preserve. Such permits are expensive and time-consuming to acquire and would still result in 80 to 90 percent of the Vaught property being placed in protective habitat custody.

"This would have left us with only two hilltops to develop, and regulations already in place would almost certainly prevent access to them," Kim said.

An appeal to reduce taxes on the property (based on houses in the area valued in the $500,000 range) was also turned down.

"The Goldencheek's squatters' rights have resulted in parasitism on us—not by the Brown-headed Cowbird but by the Balcones Canyonlands Preserve," Kim said. The Vaughts still want to sell their land.

It is sad indeed to look back to the old days and recall Warren Pulich's futile efforts in 1973 to get a mere $2,000 to make a full census of the Golden-cheeked Warbler. In connection with the vast brush-cutting program begun in the 1950s by the Department of Agriculture, Pulich reported to the Office of Endangered Species of the U.S. Fish and Wildlife Service that ten counties had suffered a total loss of Goldencheek habitat, leaving the bird's 1973 range at about thirty counties. (The present range is about twenty-four counties, but many of those have little or no recent data.)

Frustrated, Pulich was forecasting what proved to be all too true: "It would be cheaper and easier to keep the Golden-cheek from becoming acutely endangered now than it would be to effect its recovery later."

In 1993 the four-year cost in money, human efforts, and passion to save the Golden-cheeked Warbler, the Black-capped Vireo, and the other endangered species was figured to run $164.8 million. The truth is that with the many people working in organizations, both public and private, and the hours, which are sure to be countless, the true costs of the full recovery efforts are incalculable.

The struggle to save parts of the earth as wild will continue, and the question of "environmental morality" will become ever more intense. I choose the word "morality" rather than "ethics" because "morality" applies to personal behavior more than "ethics," and in the end what matters will be how individuals believe and act. That belief will then become what humans as a whole will believe and do. Only knowledge and will can save a healthy environment. The major question is, how much time do we have to get smart?

It is a given that the value of wilderness preserved will increase with its rapidly growing scarcity. I hope (how can I dare predict?) that our great-great-great-grandchildren, having their own wilderness experience, will look at what we have preserved and marvel, "How did they get so much so cheap?" Then they undoubtedly will ask, "Why didn't they get more?"

***

"We now have the technology and expertise to identify and quantify winter habitat for the Golden-cheeked Warbler," John Rappole, research scientist at the Conservation and Research Center of the Smithsonian Institution, reported to me in October 1998. "We have used this expertise to create an accurate map of warbler distribution in a 31,000-square-kilometer area in southwestern Honduras, and the same expertise could be used to create a map for the entire winter distribution of the species."

I had lived at least thirty years with the mystery of where the Goldencheek spends its winters, so for me this was extraordinary news.

Having always chased birds on foot, I had to make an Olympic leap from footsteps to Rappole's satellites and computers. With his report in my hand, I was also promptly faced with hard reality: I am a nonscientist, and I also happen to be a nonagenarian who can clearly recall my rides in a horse-drawn buggy, grocery deliveries by wagon, and my first sight of an airplane as it barnstormed in the sky above a hay field. To understand Rappole's complex research, I obviously needed help—a lot of it.

I called Jim Dick, who does mapping and cartography for the U.S. Fish and Wildlife Service. He told me Rappole's report was necessarily detailed so that future scientists could rely on it, and he bolstered my confidence by describing some work methods and defining words and acronyms that weren't in my dictionary.

I went back to the report with solid determination. I *would* understand it thoroughly, and I *would* write about it in a way that any other ninety-one-year-old nonscientist could understand it.

For his study Rappole had the cooperation of the Honduran Forest Service, the Honduran Ministry of the Environment, the Inter-American Development Bank, the World Bank, and Professor Sherry Thorn of the Department of Biology at the University of Honduras, who acted as the principal Honduran collaborator.

Rappole had for his field team leader Dave King, a doctoral student at the University of Massachusetts, who was assisted by José Thorn and some of Professor Thorn's students. Rappole also contracted with Doug Muchoney, who was pursuing a doctorate in the spatial analysis program at Boston University, to provide an unsupervised classification (one that has not been ground-proofed) of the Landsat TM (Thematic Mapper) scene covering the principal known localities of the Goldencheek in Honduras.

The program included the use of Landsat TM scenes for intensive field-work incorporating remote sensing, GIS (Geographic Information System), and field investigation to identify and prioritize the principal winter habitats for the Golden-cheeked Warbler within the area covered by the Landsat scene in the country of Honduras.

A Landsat TM scene is a digital picture of the earth made from a satellite. It is roughly 180 km on a side and thus covers about 10,000 square miles. Rappole originally had two such scenes for Honduras. One was for a region covering the western portion of the state of Olancho, and one was for the area including Cusuco National Park near the city of San Pedro Sula on the Guatemalan border.

I studied John Rappole's report. I solved one problem, and then the next. Time passed. The telephone reference lady at the library helped with another. Time marched on. Finally, I called John and told him I was writing what I could, and I would send it to him for corrections and improvements. "But please, John," I said, "leave it so that old-timers can easily understand it." The following is what we came up with together.

* * *

Though high-tech, the project began the old-fashioned way—on foot. Based on sightings and specimens, the habitat for the Goldencheek in Honduras is pine-oak forests above one thousand meters elevation. Likely locations were identified by maps, aerial photos, Honduran Forest Service habitat maps, and personal contacts.

With this information, observers began by walking through likely-looking habitat, listening for vocal members of the mixed-species flocks frequented by Goldencheeks. The observers then stayed with such a flock until a Golden-cheek was found, or they were convinced it contained no Goldencheeks. The average time required for this type of endeavor was about two hours. For each warbler sighted, date, time, sex, band presence, number of associated conspe-cifics, number and species of flock associates, and foraging tree species were recorded.

Next, using a global positioning system (GPS), the location of every war-bler sighting was then registered for altitude, longitude, and latitude, and a vegetation analysis using canopy height, dominant tree species, tree dbh (diameter at breast height), shrub density, canopy cover, and ground cover was also recorded. By the use of stratified transects in appropriate habitat, the frequency of flock occurrence was also determined.

Rappole joined the field team in Tegucigalpa and proceeded to assess the accuracy of these Landsat scenes by visiting several hundred land sites on a thousand-mile journey covering much of central and western Honduras. Now with the satellite scenes checked and conforming with the ground records, and with the field data completed by the team headed by Dave King, it was possible to begin analysis.

Peter Leimgruber and Tim Boucher of the Smithsonian Conservation and Research Center's Spatial Analysis Lab were contracted to obtain the Siguatepeque scene, which covers most of the known Golden-cheeked Warbler locations in southwestern Honduras. It consists of seven spectral bands that range from the visible spectrum to the far infrared. All seven were systematically corrected, projected, and rectified.

This scene was then divided into classes, each of which represented a subject such as pine-oak forest or water or land below the one-thousand-meter elevation.

The Digital Chart of the World (DCW) was originally developed for the U.S. Defense Mapping Agency, and Rappole's study used the chart to delineate such things as roads, towns, and political boundaries. The DCW also provided elevation data.

Georeferencing is a process used to assign map coordinates to an image. The system was refined by cross-referencing and resampling to achieve a root mean square error below 1.0.

Computers were "trained" to do various stages of refinement, with the final outcome the ability to produce and divide information into categories and subcategories.

A raster is the scan pattern in which an area is scanned side to side in lines from top to bottom, a pattern of closely spaced rows of dots to form an image on a television or computer screen. A raster could block out all land below the one-thousand-meter altitude.

Pixels are small discrete elements that together constitute an image (as in the pixels in your television screen). Computers were "trained" to recognize patterns in the data and to assign pixels to the appropriate categories.

Ground truth data were registered using GPS. In addition, stereoscopic aerial photography was used for two locations, including twelve aerial photos.

Preliminary tests of the classification indicated it had an accuracy of roughly 70 percent (i.e., 70 percent of pixels assigned to a specific habitat are actually that habitat type). On a later trip the classification was improved by collecting corrected GPS locations at fifty randomly selected points for each major habitat type.

By combining the various layers of information, Rappole was able to come up with the final product—a color map showing the Golden-cheeked Warbler's winter habitat. At long last, here was the answer to a puzzle of many decades, truly a remarkable accomplishment.

***

Of course, studies like Rappole's also come up with other observations, such as the fact that foraging behavior of the Goldencheek is quite distinct. The species is generally slower and more methodical in searching for food than similar related species that often occur as flock associates: the Black-throated Green Warbler, Townsend's Warbler, and the Hermit Warbler. Goldencheeks make short, brief flights from one portion of the canopy to another, and they spend several seconds carefully searching the clumps of leaves in a particular area before flitting to another. This behavior contrasts sharply with the more "nervous" movements of both Black-throated Green and Townsend's Warblers. Hermit Warblers were seldom observed to forage in substrate other than pine.

It is estimated that 5 to 10 percent of the Goldencheek population is banded. Of the seventy-seven birds observed, it was determined that sixty-eight of them wore no bands, and for eight birds no clear view of the leg could be obtained. King thought that for one bird he *might* have seen a band. Given the number of birds observed, it was thought that at least a few banded birds should have been found. Two explanations were proposed: (a) given that the majority of banded birds derive from Fort Hood, Texas, then the birds from Fort Hood do not winter in Honduras; or (b) the size of the breeding populations has been underestimated.

The seventy-seven Goldencheeks sighted in the study were found at sixty-four distinct localities, and, as expected, they were found in pine-oak forest above an elevation of 1,000 meters. The lowest elevation at record of which Rappole was aware was reported by a Peace Corps worker, Mark Bonta, at 1,000 meters in La Tigra National Park. (Red and I sighted a male and a female at 2,900 feet at Laguna Bélgica, Chiapas, in 1992.) The highest elevation at which the study found the bird was at 2,350 meters, just outside the town of Guajiquiro.

As in Texas, the Goldencheek was found among oak trees. Rappole also offered the supposition that the recent sightings of the Goldencheek in Mexico during the winter months (November–January) where none had been reported in the past could indicate that total prime habitat has dwindled and the birds are being forced into less desirable regions.

In fact, Rappole concluded that the winter range is restricted to a relatively small number of localities in Honduras, Guatemala, Mexico, and perhaps

Nicaragua. Thus the Goldencheek's population (and survival) may be limited in the winter by the availability of pine-oak habitat. Maps that show what a meager amount of suitable habitat remains are sobering.

I found there was more for me to say about John Rappole's research than the description above. Perhaps it was being untutored in the field that made me marvel at the research so much, but an odd thing happened to me while I was studying and struggling to interpret it for myself and other nonscientists.

I read the report in small parts. I studied it with interest and sincerity and real effort, and then everything stopped. I don't know how I got there, but, whatever the process, I had reached a state where I was awed and enthralled. This makes a poor rhyme, and it is an even more impoverished description of my mental and emotional state, which bordered on the giddy delightful.

Did I begin to think in layers somewhat like those computers the researchers trained? It was one of those times when one not only recognizes a large truth but also experiences deep emotion about it.

Perhaps my thoughts began incubating after a conversation I had with Peter Stangel, director of the Neotropical Migratory Bird Conservation Initiative at Partners in Flight.

Back in 1994 Peter had told me that Partners in Flight had grown to include 119 government and private agencies in the United States, and Aves de las Americas had programs designed for the benefit of neotropical migrants in twenty countries.

Now Peter was telling me about developments connected specifically with the Golden-cheeked Warbler. "Marjorie," he said, "as of 1998 your special bird has gained quite a bit of notoriety. It's because of work and effort put in by many people like you and Red and a large partnership of groups devoting time and money specifically to the warbler."

He began to name some of them off: "There's at least a dozen or more . . . the Nature Conservancy, the National Audubon Society, the Bureau of Land Management, the U.S. Fish and Wildlife Service, Partners in Flight in Central America and Mexico, the International Association of Fish and Wildlife Agencies, the U.S. Geological Survey, the Department of Defense . . ."

"The Department of Defense?" I started to question, but then I remembered the stack of documents four inches high that John Cornelius, endangered species program director at Fort Hood, Texas, had sent me. This military installation has approximately 40,110 acres of Golden-cheeked Warbler habitat with a population of perhaps more than two thousand birds.

And just take into account the changes in Partners in Flight itself. It began in 1990 as a mostly volunteer organization with biologists and other scientists in various organizations taking on, largely without pay, extra duties that were connected with migratory birds. Now Andy Romero, coordinator of programs for Partners in Flight, tells me the organization has salaried coordinators in charge of five different regions in the United States. Each region, of course, includes many people.

"Oh, yes," Cliff Shackelford, of Partners in Flight at the Texas Parks and Wildlife Department, agreed. "You'd be surprised. A lot of schools and students are doing research on the Goldencheek. This includes people from all around the country. Just last week a fellow from William and Mary is planning a thesis."

I hadn't even begun thinking about corporations: They have money. They pitch in and help. There are lots of 'em.

I closed my eyes and let the pixels of my mind whirl around all these people who are thinking and planning and spending time, energy, and money for the sake of a teeny bird.

I just let my awe and enthrallment carry me along somewhat like a trap-door spider balloons along on a skein of silk lazily circling in a breeze, and, as is so often the case at my age, I was swept back to comparing the present with the past.

The vision started right at the beginning, with Osbert Salvin traveling on horseback in a highland forest in Guatemala in 1859 and stopping to shoot a bird so he could hold it in his hand. In his day almost no one was interested in birds other than eating them. Then — *whoosh!* — here is John Rappole's team working in similar highland pine-oak forests of the tropics, and I'm remembering well how steep the inclines can be in those mountains and how slippery and treacherous the pine needles are underfoot. I remember, too, how flocks can pass through the groves in seconds, and I could understand the effort it must have taken for the observers to keep up with them. This is the way fieldwork was done for decades.

Leap to the present, where a system can find and record the altitude and the exact geographic position in the whole wide world where a single small bird is sighted.

Great Jumpin' Jehosaphat! It's overwhelming!

Some years ago, riding a raft in a swift river rapids, I was totally terrified for a short interval, and later, when I thought about my fear, I decided it is good for us every once in a while to be reminded we are mortal.

Just as important, it seems, would be to pause every once in awhile to realize what debtors we are to the millions who have gone before us and who have spent countless hours of work and thought that led to the invention and refinement of such things as binoculars, telescopes, cameras, computers, and satellites and other inventions that have enabled John Rappole and his several

Celebrating the one-hundredth anniversary of the national wildlife refuge system are (left to
right) *Congressman Lamar Smith; retired congressman Jake Pickle; Marjorie Adams; and
Debra Holle, Balcones Canyonlands National Wildlife Refuge manager. Marjorie is saying,
"Thanks, Jake, for the money you helped raise for the refuge." Jake is saying, "I finally had
to tell Marj that warblers don't vote." Lamar is saying, "My wife and I went on our first
birding trip last week, and we tallied fifty birds." Debra is thinking, "This is a wonderful
one-hundredth birthday party for the national wildlife refuge system." All are saying,
"Let's wish for another hundred years." (Courtesy of Clifton Ladd)*

teams, working in two countries, to produce a color map showing where an
endangered bird lives in the winter.

But no matter how awed or enthralled I was, in the real world there remains the question, can all the king's horses and all the king's men, putting
all their concerted effort together, actually save the Golden-cheeked Warbler?

I had hardly written these words when Nature sent at least one answer. Listen to the news. Twenty-five inches of rain in four days! Dear Heaven! Look
at those before-and-after aerial views of Honduras on the television, revealing
that an entire country has been scoured.

Hurricane Mitch had left thousands of people dead and hundreds missing,
lives left in direst distress. When will there ever be time, thought, or energy
to give to a small bird?

In the environmental world we have long said that the greatest sin is to give up hope. I had begun sinning (forgive me—it was only on the blackest days) before Mitch struck. Now no more sinning, not even an instant. Work, teach, solve, earn, demand, and do what must be done.

＊＊＊

We have learned that it should be safe again to film in Mexico. Laguna Bélgica awaits.

Ah me! Ah-h-h-h me! You're hearing an old lady sighing.

Ah me! And she bows her head. She sits silent as if in prayer; then suddenly there's an explosion. Surely you must know that, given certain circumstances, even the gentlest old ladies can be reduced to cussin'.

"Damn this old age! Damn! Double damn! A thousand curses on its invention!" But damn and cuss as she might, there is no cure.

The assessments of our age and abilities made in 1992 by our son and daughter have been proven by time. Red and Marjorie Adams now are factually, actually, incontrovertibly too old to head south, a l-o-n-g way south, to film our beloved Goldencheek in its winter home.

The old lady sits, quietly dreaming old dreams. We were there. We found them. Surely they're still there. Remember the message: Don't give up hope. Just revise if necessary.

The old lady turns, and she's looking straight at you as you read this. Her gaze is piercing and commanding. You. Yes, *you.* Over there in the corner. And you, sitting in your office. You, leading the school kids on a field trip. *You* with the camera. Can *you* go? Can *you* volunteer?

Can you hear us?

EPILOGUE

It's a Long Way from There to Here

When Mary Marjorie Valentine was born December 7, 1913, in San Antonio, Texas, the doctors laid her aside for dead. Well, I fooled 'em, and I've lived on borrowed time now for more than ninety-one years. Achieving genuine old-timer status means I can look back much more than forward, and as I do some reliving, I conclude that my generation derived a huge bonus by living in a time of truly extraordinary change. Consider the following conditions when I was born in 1913.

The president of the United States was Woodrow Wilson, a highly educated man seeking world peace and law.

The Panama Canal locks were recently completed.

The silk strike laid down the foundation for workers' rights in America.

The Department of Labor became a cabinet office.

The average annual income of an employee was $640.

The Victorian age in America was ending.

Rabindranath Tagore won the Nobel Prize for literature. (He also gave Gandhi the title of "mahatma.")

Marcel Duchamp's notorious painting, *Nude Descending a Staircase,* was the runaway hit of the 1913 Armory Show (the first major showing of avant-garde art in the United States).

Poetry magazine published Joyce Kilmer's "Trees."

The Sixteenth Amendment, providing income taxes on individuals and corporations, was passed.

The Seventeenth Amendment, providing direct election of senators by the people, went into effect.

A black man could vote. A woman, white or black, could not.

Men owned property, including their wives' inheritances—yes, even jewels handed down from her grandmother.

A woman's place was in the home.

Sex was not a topic of conversation.

There was no planned parenthood.

Only the chosen few attended college.

In the South a white might whip a black servant, and no one would report it.

Street cleaners swept up after horses.

Railroads were the major transportation from town to town.

The Model T Ford was less than five years old.

Less than 10 percent of the nation's highways were usable by automobiles.

It was not unusual for travelers in rural areas to ask for overnight shelter at a stranger's house.

Flying machines sat in fools' barns, waiting to fly.

There were 9,513 telephones in the United States. Customers were doctors, certain industries, and the very rich.

Camel cigarettes came on the market.

The zipper was patented.

A one-room schoolhouse with one ceiling-to-floor window cost about $2,000.

Most of the populace lived in rural areas. Outdoor toilets were the norm.

There were no child labor laws.

Children died of whooping cough, diphtheria, and measles and during crop failures.

Adults hoped to survive measles, chicken pox, smallpox, tuberculosis, pneumonia, and childbirth.

A telegram usually meant a death.

The average life span was fifty-four years.

Many people didn't lock their doors at night.

People knew and helped their neighbors.

And in the natural world of 1913 . . .

The planet Pluto was yet to be discovered.

Two Danish astronomers proved the existence of star systems outside the Milky Way.

Albert Einstein began his work on the general theory of relativity.

The Weeks-McLean Act established federal control over migratory birds, then was declared unconstitutional.

No federal law prohibited the sale or exportation of game.

Strychnine-soaked bran or corn was spread over fields to kill grasshoppers and blackbirds.

There were thirty-nine sparsely membered state Audubon societies. One state boasted two, neither of which knew the other existed.

Twenty-seven states required some form of hunting license.

With or without laws, hunters killed almost anything in any season in any numbers.

Automatic guns with the capacity to shoot five times before reloading were popular with hunters.

In North Dakota it was illegal to use automobiles to hunt wildfowl.

Cedar birds (Cedar Waxwings), grosbeaks, larks, blackbirds, grackles, and doves were listed as game birds.

In southern states, tens of thousands of robins were killed for food.

There remained in some states a closed season on the Passenger Pigeon.

The term "game hog" (for a hunter who killed more than the limit or only for sport) had already been invented.

It was a "known fact" that forty-three species of birds feed on the cotton boll weevil.

The National Geographic Society published its first bird book.

The latest report on the number of bird species in the world was in the 1910 volume of *Encyclopedia Britannica:* 10,300 species.

Women wore aigrettes and plumes on hats and gowns.

Collecting eggs was both a hobby and a business.

Most people in the United States had never seen a pair of binoculars.

William Hornaday, director of the New York Zoological Park, was championing a resolution by Congress to have all national forests, forest reserves, and parks declared game reserves.

The same gentleman also said the Whooping Crane would almost certainly be "totally exterminated."

Many people in the world relied on wood as their sole source of fuel.

People filled swamps and estuaries because they caused malaria.

One of the first things settlers did was set fire to the land to clear it of everything.

There was always new land to move on to when the old land wore out.

The major interest in nature was what made crops grow.

At age ninety-one, I look back and decide my life was a good one because . . .

If people were bred like cattle, I would have been shipped off in the first load.

I swore off whiskey at the age of six months.

Floyd Pruett sold me a computer and survived my learning it when I was seventy-nine.

Dozens of strangers took me under their birding wings.

I did my bit to help Jim Tucker start the American Birding Association.

Vegetarian Jim Tucker graciously ate our beans without mentioning the ham
they were cooked with.

I saw birding become a national competitive sport.

When that most excellent ornithologist Pauline James told me in the kindest
way that a nonscientist shouldn't be writing about birds, it was too late.

George Miksch Sutton kissed me. I was expressing my great gratitude for his
life achievements for art, birds, and the environment.

Walter Cronkite allowed me to hug him for his narration of the outstanding
film *Pointless Pollution*. (He patted my shoulder.)

I will swear on a stack of Petersons that I saw champion birder Kenn Kaufman
open a can of hominy at the banquet of the very first convention of the
American Birding Association.

I have been critiqued by Alexander Sprunt IV.

I survived "sanitizing" by the great Edgar Kincaid Jr., known as the Old
Cassowary.

On Galveston Island, Texas, I beheld a teenaged Victor Emanuel *birding*.

I saw Ed Kutac begin his Life List—now about 550 and, as is worth noting,
500 are in Texas.

The wonderful organization, Partners in Flight, is run mostly by volunteers.

It's true: Andre Sada's binoculars are trained to find the best birds—and not
just in Mexico.

I ate a Brown-headed Cowbird. Not bad. Recommendation: national contest
for best recipe.

I ate a roadkill. (Saw the car ahead hit the Northern Bobwhite. Not a mark
on it.)

At the Christmas Countdown, I survived the discovery that I had a Mourning
Dove in my lunch.

Peter Matthiessen gave me one of his secrets for staying well in foreign lands:
Begin with small amounts of native food, and increase gradually.

When I mentioned that Red had been president of our local Audubon Society
in Texas, the guard opened the gate and let us proceed to Gallon Jug,
Belize.

Charles Hartshorne autographed *Born to Sing* for me when he was a hundred
years old.

I saw Olga Clarke survive finding the Connecticut Warbler in Canada.

I knew Arnold Small before Edgar Kincaid christened him the Great Buddha.

Chuck Bernstein trusted me to correct what we thought was a myna bird.

Pete Dunne forgave me for meeting him in the wrong place.

Edgar Kincaid Jr. didn't always wear a tie when he went birding.

I petted a baby ostrich. (Not at all soft and fluffy.)

My column, Bird World, was the tool that started a new nature club in Wichita Falls, Texas, and another one in Victoria, Texas.

A reader who had just met me told me she loved me.

I have considerable more sense now than when I shot dead, with a .22 rifle, a flying American Crow.

I had an eye-to-eye séance with a Keel-billed Toucan. He learned the most.

By adapting my birding skills, I persuaded Mollie Ivins's roaming young poodle to come to me. I lowered my head and meowed like a cat.

State and national parks now have bird checklists.

Someday people will be able to just take off and fly.

The organization Trees, Water, and People is trying to get chimney stoves that use up to 70 percent less wood into households in Central America. At $60 each they are still too expensive for most households. (A solar oven that costs less than $4 has not been as successful as hoped. It doesn't give a smoke taste to food.)

I still get almost as much pleasure learning something new about an old bird as I do seeing a new one.

I lived to see birding become an international industry.

The Great Texas Birding Classic added an event called the Outta-Sight Song Birder Tournament for blind birders.

The National Fish and Wildlife Foundation was born.

Partners in Flight, also called Aves de las Americas, is now American Bird Conservancy for the whole Western Hemisphere.

My grandchildren are proud of their bird names. So far they are pleased with my bird names for their children.

These kids won't be too disappointed I'm not taking them to the moon for the weekend. No birds there.

Instead we will have enlightening discussions about harnessing the vast energy in the empty space of the universe.

New bird species are still being discovered.

Some people are now feeling Mother Earth's pulse.

I think I'm learning the elusiveness of truth.

I had the Golden-cheeked Warbler as a teacher.

I can still see. I can remember some field marks.

Despite having broken my wrists, I can hold lightweight binoculars.

With a hearing aid, I can hear a bird.

I have been married to the same man for seventy-three years, and I will still count on him to identify that birdsong for me.

I keep busy, busy, to finish all the things on my list. (Who should I leave the list to?)

It's morning. I just woke up.
I have been given another day!
Yes, Red Adams, this is abso-goshdurn-lutely wonderful!

Lightning Source UK Ltd.
Milton Keynes UK
05 December 2010

163892UK00001BA/4/P